Holt Advanced Spanish

Curso dos

Lesson Planner

with Differentiated Instruction

HOLT, RINEHART AND WINSTON

A Harcourt Education Company

Orlando • **Austin** • New York • San Diego • Toronto • London

Contributing Writer

Argelia Guadarrama

Printed in the United States of America

ISBN 0-03-039777-4

1 2 3 4 5 6 7 082 06 05 04

Table of Contents

50-Minute Lesson Plans

To the Teacher

You will find that the ***Nuevas vistas*** *Lesson Planner* facilitates the planning, execution, and documentation of your classroom work.

For more help in organizing your lessons, use the *One-Stop Planner* CD-ROM. For each chapter, this computer-ready version of the *Lesson Planner* includes: editable lesson plans with direct links to teaching resources, printable worksheets from resource books, direct launches to the HRW Internet activities, video and audio segments, the *Test Generator,* and clip art for vocabulary items.

Standards for Foreign Language Learning

At the top of each page of the lesson plan is a box showing the correlations of the *Student Edition* and the *Teacher's Resource Binder* to the Standards for Foreign Language Learning. In addition, the chart on page viii provides a summary of the five major goals and the standards within each goal, so that you can more easily understand the correlation within each section of the *Lesson Planner*.

Calendario de tareas cotidianas

On page vi you will find a homework calendar (**Calendario de tareas cotidianas**) that you can copy and distribute to your class so that you and your students can keep track of their assignments each week.

Collection-Specific Lesson Plans

Each daily lesson plan consists of timed suggestions for core instruction (regular lesson plans total 50 minutes), Optional Resources (also timed) that can be integrated into the lesson plan, and correlations to the Standards for Foreign Language Learning. There are 30 50-minute lesson plans per collection.

Differentiated Instruction

Within each collection's lesson plans are suggestions for differentiating instruction.

- ▲ Advanced Learners
- ◆ Slower Pace Learners
- ● Students with Special Learning Needs
- ■ Heritage Speakers

Differentiated Instruction

Carol Ann Tomlinson and Cindy Strickland

Teachers who differentiate their instruction recognize that students are at different points in their learning journeys, will grow at different rates, and will need different kinds and amounts of support to reach their goals.

Differentiation and Varied Approaches

Differentiated classrooms offer varied approaches to **content** (what students learn), **process** (how students go about making sense of essential knowledge and practicing essential skills), **product** (how students demonstrate what they have learned), and **learning environment** (the setting in which students learn). Differentiation is based on an ongoing diagnosis of student interest, learning profile, and readiness.

Differentiation and the World-Language Teacher

World-language teachers are natural differentiators for learning profile. We provide opportunities for students to acquire proficiency in the target language through a variety of means: speaking, listening, writing, and reading. Through this variety of approaches, we recognize that students' proficiency in each of these skill areas will vary. Good language teachers work hard to help students improve in areas in which they struggle, and revel in areas of strength.

Systematic differentiation for readiness provides many world-language teachers with a bit more of a challenge. Students come to us with a huge range in amount and type of language experience, including, for example, first-year students who have had no exposure to the target language, who have had an exploratory class, who have studied another target language, or who are native speakers.

Key Principles of Differentiated Instruction

There are several key principles to follow when differentiating instruction in the language classroom. First, start by clearly defining what is most essential for students to know, understand, and be able to do in the target language. Second, hold high expectations for all students and make sure that they are engaged in **respectful work**. Third, use **flexible grouping**, an excellent tool to ensure that all students learn to work independently, cooperatively, and collaboratively in a variety of settings and with a variety of peers.

A final principle of differentiated instruction is **ongoing assessment**. To this end, the teacher constantly monitors student interest, learning profile, and readiness in order to adjust to the growing and changing learner. Teachers must not assume that a student will have the same readiness or interest in every unit of study or in every skill area. Preassessment is a must, particularly in the areas of knowledge and facility with vocabulary and grammatical constructions.

The Role of the Teacher in Academically Diverse Classrooms

Good teachers have always recognized that "one size fits all" instruction does not serve students well. To be effective, teachers must find ways consistently to **reach more kinds of learners more often**—by recognizing and responding to students' varied readiness levels, by honoring their diverse interests, and by understanding their preferences for how they learn information and practice new skills.

Differentiated Instruction Bibliography

The following references are a starting point for the teacher interested in learning more about the topic of differentiated instruction.

Gardner, H. *Multiple Intelligences: The Theory in Practice*. New York: Basic Books, 1993.

Tomlinson, C. *How to Differentiate Instruction in Mixed-Ability Classrooms*, 2/e. Alexandria, VA: Association for Supervision and Curriculum Development, 2001.

Udall, A., and Daniels, J. *Creating the Thoughtful Classroom: Strategies to Promote Student Thinking*. Tucson, AZ: Zephyr Press, 1991.

Winebrenner, S. *Teaching Kids with Learning Difficulties in the Regular Classroom*. Minneapolis, MN: Free Spirit, 1996.

Calendario de tareas cotidianas

Fechas: del **lunes** ______________________ al **viernes** ______________________

Día	Tareas
lunes, el ____	*Textbook:* ____ ***Cuaderno de práctica:*** ____ *Advanced Placement Literature Preparation Program:* ____ *Other:* ____
martes, el ____	*Textbook:* ____ ***Cuaderno de práctica:*** ____ *Advanced Placement Literature Preparation Program:* ____ *Other:* ____
miércoles, el ____	*Textbook:* ____ ***Cuaderno de práctica:*** ____ *Advanced Placement Literature Preparation Program:* ____ *Other:* ____
jueves, el ____	*Textbook:* ____ ***Cuaderno de práctica:*** ____ *Advanced Placement Literature Preparation Program:* ____ *Other:* ____
viernes, el ____	*Textbook:* ____ ***Cuaderno de práctica:*** ____ *Advanced Placement Literature Preparation Program:* ____ *Other:* ____

STUDENT PROGRESS REPORT

Name ______________________ Class ______________ Date ______________

From ______________ To ______________

	1 Very poor	2 Poor	3 Average	4 Good	5 Excellent	Comments
Reading						
Speaking						
Writing						
Quizzes/Tests						
Completion of assignments						
Class participation						

Overall effort and involvement in language learning

____

____

____

Teacher signature ______________________

Parent or guardian signature ______________________

STANDARDS FOR FOREIGN LANGUAGE LEARNING

Communication Communicate in Languages Other Than English	**Standard 1.1**	Students engage in conversations, provide and obtain information, express feelings and emotions, and exchange opinions.
	Standard 1.2	Students understand and interpret written and spoken language on a variety of topics.
	Standard 1.3	Students present information, concepts, and ideas to an audience of listeners or readers on a variety of topics.
Cultures Gain Knowledge and Understanding of Other Cultures	**Standard 2.1**	Students demonstrate an understanding of the relationship between the practices and perspectives of the culture studied.
	Standard 2.2	Students demonstrate an understanding of the relationship between the products and perspectives of the culture studied.
Connections Connect with Other Disciplines and Acquire Information	**Standard 3.1**	Students reinforce and further their knowledge of other disciplines through the foreign language.
	Standard 3.2	Students acquire information and recognize the distinctive viewpoints that are only available through the foreign language and its cultures.
Comparisons Develop Insight into the Nature of Language and Culture	**Standard 4.1**	Students demonstrate understanding of the nature of language through comparisons of the language studied and their own.
	Standard 4.2	Students demonstrate understanding of the concept of culture through comparisons of the cultures studied and their own.
Communities Participate in Multilingual Communities at Home and Around the World	**Standard 5.1**	Students use the language both within and beyond the school setting.
	Standard 5.2	Students show evidence of becoming life-long learners by using the language for personal enjoyment and enrichment.

50-Minute Daily Lesson Plans

Teacher's Name ______________________ Class ______________________ Date ______________

COLECCIÓN 1

Esfuerzos heroicos

DAY 1 50-MINUTE LESSON PLAN

STANDARDS FOR FOREIGN LANGUAGE LEARNING: DAY 1

Lectura

Communication 1.1: Students engage in conversations, provide and obtain information, express feelings and emotions, and exchange opinions.

Communication 1.2: Students understand and interpret written and spoken language on a variety of topics.

Communication 1.3: Students present information, concepts, and ideas to an audience of listeners or readers on a variety of topics.

Connections 3.1: Students reinforce and further their knowledge of other disciplines through the foreign language.

Communities 5.1: Students use the language both within and beyond the school setting.

CORE INSTRUCTION

Warm-Up

- (5 min.) Have students read the objectives in the Collection Opener, p. xxii. See Collection Overview, *TRB,* p. 4.

Lectura

Teach

- (5 min.) Have students do **Antes de leer: Punto de partida,** p. 2. See **Punto de partida,** *TRB,* p. 5.
- (10 min.) Have students discuss **Telón de fondo,** p. 3. See **Telón de fondo,** *TRB,* p. 5.
- (10 min.) Have students read aloud **Conoce al escritor,** p. 10.
- (15 min.) Have students begin reading aloud **de *Autobiografía de un esclavo,*** pp. 4–9.

Wrap-Up

- (5 min.) Discuss **Elementos de literatura: El género biográfico,** p. 42. See **Elementos de literatura,** *TRB,* p. 18.

OPTIONAL RESOURCES

- (10 min.) See Presentation Suggestions, item one, *TRB,* p. 4.
- (10 min.) See Presentation Suggestions, item two, *TRB,* p. 4.
- (10 min.) See Presentation Suggestions, item three, *TRB,* p. 4.
- (10 min.) See **Vocabulario en contexto,** Group Work, *TRB,* p. 23.
- (10 min.) Read aloud **de *Autobiografía de un esclavo,*** Summary, *TRB,* p. 5. ◆ ●
- (10 min.) Have students do Getting Started, *TRB,* p. 18.

Practice Options/Homework Suggestions

- Internet (go.hrw.com, Keyword: WN3 ESFUERZOS-LEC)
- Have students finish reading **de *Autobiografía de un esclavo,*** pp. 4–9.
- Have students practice the Reading Strategy, *TRB,* pp. 234–235. ◆ ●
- Have students study **Vocabulario adicional, de *Autobiografía de un esclavo,*** *TRB,* p. 297. ◆ ●
- *Advanced Placement Literature Preparation Book,* pp. 1–5 ▲

▲ = Advanced Learners ◆ = Slower Pace Learners ● = Special Learning Needs ■ = Heritage Speakers

Esfuerzos heroicos

DAY 2 50-MINUTE LESSON PLAN

STANDARDS FOR FOREIGN LANGUAGE LEARNING: DAY 2

Lectura/Vocabulario

Communication 1.1: Students engage in conversations, provide and obtain information, express feelings and emotions, and exchange opinions.

Communication 1.2: Students understand and interpret written and spoken language on a variety of topics.

Connections 3.1: Students reinforce and further their knowledge of other disciplines through the foreign language.

Comparisons 4.1: Students demonstrate understanding of the nature of language through comparisons of the language studied and their own.

Communities 5.1: Students use the language both within and beyond the school setting.

CORE INSTRUCTION

Warm-Up

- (5 min.) Have students read **Elementos de literatura: ¡Extra! ¡Lee las últimas noticias!** and **Hecho y opinión,** p. 42.

Lectura/Vocabulario

Teach

- (20 min.) Have students read aloud **de *Autobiografía de un esclavo,*** pp. 4–9.
- (15 min.) Have students do **Vocabulario en contexto,** Activity A, p. 47. See Activity A, *TRB,* p. 23.

Wrap-Up

- (10 min.) Have students do Applying the Element, *TRB,* p. 18.

OPTIONAL RESOURCES

- (20 min.) See Techniques for Handling the Reading, *TRB,* p. 6. ◆ ●
- (10 min.) See Language Note, *TRB,* p. 18.
- (30 min.) Have students do Art Link, *TRB,* p. 19.
- (15 min.) Have students do Literature Link, *TRB,* pp. 18–19.

Practice Options/Homework Suggestions

- Internet (go.hrw.com, Keyword: WN3 ESFUERZOS-LEC)
- Have students study **Vocabulario esencial, de *Autobiografía de un esclavo,*** p. 73. ◆ ●
- *Cuaderno de práctica,* Activities 1–4, pp. 1–2

▲ = Advanced Learners ◆ = Slower Pace Learners ● = Special Learning Needs ■ = Heritage Speakers

Teacher's Name ______________________ Class ______________________ Date ____________

COLECCIÓN 1

Esfuerzos heroicos

DAY 3 50-MINUTE LESSON PLAN

STANDARDS FOR FOREIGN LANGUAGE LEARNING: DAY 3

Lectura

Communication 1.1: Students engage in conversations, provide and obtain information, express feelings and emotions, and exchange opinions.

Communication 1.2: Students understand and interpret written and spoken language on a variety of topics.

Communication 1.3: Students present information, concepts, and ideas to an audience of listeners or readers on a variety of topics.

Connections 3.1: Students reinforce and further their knowledge of other disciplines through the foreign language.

Communities 5.1: Students use the language both within and beyond the school setting.

CORE INSTRUCTION

Warm-Up

- (5 min.) Have pairs of students discuss **Primeras impresiones,** p. 12, using the expressions in **Así se dice,** p. 12.

Lectura

Teach

- (15 min.) Have students do **Crea significados: Interpretaciones del texto** using the expressions in **Así se dice,** p. 12. See **Interpretaciones del texto,** *TRB,* p. 6.
- (25 min.) Have students do **Investigación y exposición oral,** p. 13. See **Investigación y exposición oral,** *TRB,* p. 7.

Wrap-Up

- (5 min.) Have students do **Repaso del texto,** p. 12. See **Repaso del texto,** *TRB,* p. 6. Discuss answers.

OPTIONAL RESOURCES

- (20 min.) Have students answer questions in **Conexiones con el texto,** p. 12.
- (20 min.) Have students do **Cuaderno del escritor,** p. 13. See **Cuaderno del escritor,** *TRB,* p. 7.
- (20 min.) Have students do **Para hispanohablantes,** *TRB,* p. 7. ■
- (20 min.) Have students do **Para angloparlantes,** *TRB,* p. 7.
- (10 min.) Have students discuss **¿Te acuerdas?** and create other examples, p. 12.

Practice Options/Homework Suggestions

- Internet (go.hrw.com, Keyword: WN3 ESFUERZOS-LEC)
- Have pairs of students do **Estrategias de pensamiento,** p. 13. See **Estrategias de pensamiento,** *TRB,* p. 7. ▲
- Have students study **Vocabulario adicional, de *Autobiografía de un esclavo*,** *TRB,* p. 297. ◆ ●
- Have students practice the Reading Strategy, *TRB,* pp. 234–235. ◆ ●
- Have students study for **Prueba de lectura.**

▲ = Advanced Learners ◆ = Slower Pace Learners ● = Special Learning Needs ■ = Heritage Speakers

Teacher's Name ______________________ Class ______________________ Date ____________

COLECCIÓN

Esfuerzos heroicos

DAY 4 50-MINUTE LESSON PLAN

STANDARDS FOR FOREIGN LANGUAGE LEARNING: DAY 4

Lectura/Gramática

Communication 1.2: Students understand and interpret written and spoken language on a variety of topics.

Comparisons 4.1: Students demonstrate understanding of the nature of language through comparisons of the language studied and their own.

CORE INSTRUCTION

Warm-Up

- (5 min.) Have students review **Vocabulario esencial** for **de *Autobiografía de un esclavo,*** p. 73.

Lectura

Assess

- (30 min.) Give **Prueba de lectura: de *Autobiografía de un esclavo,*** *Assessment Program,* pp. 3–4.

Gramática

Teach

- (10 min.) Present **Los pronombres personales,** p. 52.

Wrap-Up

- (5 min.) Have students do **Los pronombres personales,** *TRB,* p. 26.

OPTIONAL RESOURCES

- (10 min.) Have students do Cooperative Learning, *TRB,* p. 26. ◆ ●
- (10 min.) Review information in Language Note, *TRB,* p. 26.
- (15 min.) Have students do **Ampliación, Hoja de práctica 1-A, Más sobre los pronombres de sujeto,** *TRB,* p. 277. ◆ ●

Practice Options/Homework Suggestions

- *Cuaderno de práctica,* Activities 1–3, pp. 12–13
- *Cuaderno de práctica,* **Ampliación, Hoja de práctica 1-A,** p. 134 ◆ ●

▲ = Advanced Learners ◆ = Slower Pace Learners ● = Special Learning Needs ■ = Heritage Speakers

Teacher's Name ______________________ Class ______________________ Date ____________

COLECCIÓN 1

Esfuerzos heroicos

DAY 5 50-MINUTE LESSON PLAN

STANDARDS FOR FOREIGN LANGUAGE LEARNING: DAY 5

Gramática

Communication 1.2: Students understand and interpret written and spoken language on a variety of topics.

Comparisons 4.1: Students demonstrate understanding of the nature of language through comparisons of the language studied and their own.

CORE INSTRUCTION

Warm-Up

- (5 min.) Have students review **Los pronombres personales,** p. 52.

Gramática

Teach

- (10 min.) Have students read **Los pronombres de sujeto** and **¡Ojo!,** p. 52. See **Los pronombres personales** and **Los pronombres de sujeto,** *TRB,* p. 26.
- (5 min.) Have students do Activity A, p. 53.
- (10 min.) Have students read **Los pronombres de complemento directo e indirecto,** pp. 53–54.
- (15 min.) Have students do Activities C, D, and E. pp. 54–55.

Wrap-Up

- (5 min.) Have students do **Los pronombres de complemento directo e indirecto,** *TRB,* p. 26.

OPTIONAL RESOURCES

- (20 min.) Have students do Cooperative Learning, *TRB,* p. 26.
- (15 min.) Have students do **Ampliación, Hoja de práctica 1-B, Variantes de la segunda persona,** *TRB,* p. 278. ◆ ●
- (15 min.) Have students do **Ampliación, Hoja de práctica 1-C, Más sobre los pronombres complementarios,** *TRB,* p. 279. ◆ ●
- (10 min.) Have students do **Para angloparlantes,** *TRB,* p. 29.

Practice Options/Homework Suggestions

- Have students do Activities B and F, pp. 54 and 55.
- Have students read **Cultura y lengua: Cuba,** pp. 14–17.
- *Cuaderno de práctica,* **Ampliación, Hoja de práctica 1-B,** p. 135 ◆ ●
- *Cuaderno de práctica,* **Ampliación, Hoja de práctica 1-C,** p. 136 ◆ ●
- *Cuaderno de práctica,* Activities 4–7, pp. 13–14
- Have students study for **Prueba de gramática.**

Assessment Options

- *Assessment Program,* **Prueba de gramática 1-A: Más sobre los pronombres de sujeto,** p. 250

▲ = Advanced Learners ◆ = Slower Pace Learners ● = Special Learning Needs ■ = Heritage Speakers

Teacher's Name ______________________ Class ____________________ Date __________

COLECCIÓN 1

Esfuerzos heroicos

DAY 6 50-MINUTE LESSON PLAN

STANDARDS FOR FOREIGN LANGUAGE LEARNING: DAY 6

Gramática/Cultura y lengua

Communication 1.1: Students engage in conversations, provide and obtain information, express feelings and emotions, and exchange opinions.

Communication 1.2: Students understand and interpret written and spoken language on a variety of topics.

Cultures 2.1: Students demonstrate an understanding of the relationship between the practices and perspectives of the culture studied.

Cultures 2.2: Students demonstrate an understanding of the relationship between the products and perspectives of the culture studied.

Connections 3.1: Students reinforce and further their knowledge of other disciplines through the foreign language.

Comparisons 4.1: Students demonstrate understanding of the nature of language through comparisons of the language studied and their own.

CORE INSTRUCTION

Warm-Up

- (5 min.) Have students review **Los pronombres personales, Los pronombres de sujeto,** and **Los pronombres de complemento directo e indirecto,** pp. 52–55.

Gramática

Assess

- (20 min.) Give **Prueba de gramática: Los pronombres de sujeto y los pronombres de complemento directo e indirecto,** *Assessment Program,* p. 13.

Cultura y lengua

Teach

- (5 min.) Present **Cultura y lengua: Cuba,** p. 14.
- (15 min.) Have students read aloud **Cultura y lengua,** pp. 14–15.

Wrap-Up

- (5 min.) Have students use the expressions in **Así se dice,** p. 18, to talk about the history and culture of Cuba.

OPTIONAL RESOURCES

- (15 min.) See **Cultura y lengua,** Teaching Suggestions, *Video Guide,* p. 3.
- (15 min.) Have students do **Hoja de actividades 2,** *Video Guide,* **Antes de ver,** p. 7.
- (10 min.) Have students do Geography Link, *TRB,* p. 8.

Practice Options/Homework Suggestions

- Internet (go.hrw.com, Keyword: WN3 ESFUERZOS-CYL)

Assessment Options

- *Assessment Program,* **Prueba de gramática 1-B: Variantes de la segunda persona,** p. 251
- *Assessment Program,* **Prueba de gramática 1-C: Más sobre los pronombres de complemento directo e indirecto,** p. 252

▲ = Advanced Learners ◆ = Slower Pace Learners ● = Special Learning Needs ■ = Heritage Speakers

Esfuerzos heroicos

STANDARDS FOR FOREIGN LANGUAGE LEARNING: DAY 7

Cultura y lengua

Cultures 2.1: Students demonstrate an understanding of the relationship between the practices and perspectives of the culture studied.

Connections 3.1: Students reinforce and further their knowledge of other disciplines through the foreign language.

Connections 3.2: Students acquire information and recognize the distinctive viewpoints that are only available through the foreign language and its cultures.

Comparisons 4.1: Students demonstrate understanding of the nature of language through comparisons of the language studied and their own.

Communities 5.2: Students show evidence of becoming life-long learners by using the language for personal enjoyment and enrichment.

CORE INSTRUCTION

Warm-Up

- (5 min.) Have students discuss History Link, *TRB,* p. 8.

Cultura y lengua

Teach

- (5 min.) Present **Así se dice,** p. 18. See **Así se dice,** *TRB,* p. 9.
- (15 min.) Have groups of students do **Actividad,** p. 18.
- (15 min.) See **Cultura y lengua,** Teaching Suggestions, **Antes de ver,** *Video Guide,* p. 7.
- (5 min.) Show **Cultura y lengua: Cuba,** *Video Program* (Videocassette 1).

Wrap-Up

- (5 min.) Play some Cuban music while discussing **Modismos y regionalismos,** p. 18. See Music Link, *TRB,* p. 8.

OPTIONAL RESOURCES

- (20 min.) Have students do Science Link, *TRB,* p. 8. ▲ ■
- (20 min.) Have students do the second Thinking Critically, *TRB,* pp. 8–9. ▲
- (10 min.) Review Language Note, *TRB,* p. 9.

Practice Options/Homework Suggestions

- Internet (go.hrw.com, Keyword: WN3 ESFUERZOS-CYL)
- Have students do Thinking Critically, *TRB,* p. 9. ▲
- Have students do Community Link, *TRB,* p. 9.

▲ = Advanced Learners ◆ = Slower Pace Learners ● = Special Learning Needs ■ = Heritage Speakers

Esfuerzos heroicos

DAY 8 50-MINUTE LESSON PLAN

STANDARDS FOR FOREIGN LANGUAGE LEARNING: DAY 8

Cultura y lengua

Cultures 2.1: Students demonstrate an understanding of the relationship between the practices and perspectives of the culture studied.

Cultures 2.2: Students demonstrate an understanding of the relationship between the products and perspectives of the culture studied.

Connections 3.1: Students reinforce and further their knowledge of other disciplines through the foreign language.

CORE INSTRUCTION

Warm-Up

- (5 min.) Have students begin reading **Cultura y lengua,** p. 16.

Cultura y lengua

Teach

- (20 min.) Have students read the paragraphs aloud, pp. 16–17.
- (20 min.) Have students do **Hoja de actividades 2, Mientras lo ves,** *Video Guide,* p. 7.

Wrap-Up

- (5 min.) Have students answer the questions in Thinking Critically, *TRB,* p. 9.

OPTIONAL RESOURCES

- (40 min.) Have students research famous Cubans who live in the United States and share one interesting fact with the class.

Practice Options/Homework Suggestions

- Internet (go.hrw.com, Keyword: WN3 ESFUERZOS-CYL)
- Have students study for **Prueba de cultura.**
- Have students read **Estrategias para leer,** pp. 19–20. ◆ ●

▲ = Advanced Learners ◆ = Slower Pace Learners ● = Special Learning Needs ■ = Heritage Speakers

Esfuerzos heroicos

DAY 9 50-MINUTE LESSON PLAN

STANDARDS FOR FOREIGN LANGUAGE LEARNING: DAY 9

Cultura y lengua/Lectura

Communication 1.2: Students understand and interpret written and spoken language on a variety of topics.

Connections 3.1: Students reinforce and further their knowledge of other disciplines through the foreign language.

CORE INSTRUCTION

Warm-Up

- (5 min.) Have students review **Cultura y lengua,** pp. 14–18.

Cultura y lengua

Assess

- (20 min.) Give **Prueba de cultura: Cuba,** *Assessment Program,* p. 17.

Lectura

Teach

- (10 min.) Have students read **Estrategias para leer,** p. 19. See Introducing the Strategy, *TRB,* p. 10.
- (10 min.) Have students do **Inténtalo tú,** p. 20. See Cooperative Learning, *TRB,* p. 10.

Wrap-Up

- (5 min.) Have students review their sentences for **Inténtalo tú,** p. 20.

OPTIONAL RESOURCES

- (10 min.) Have students do Getting Started, *TRB,* p. 10.
- (10 min.) Have students do Additional Practice, *TRB,* p. 10.
- (10 min.) Have students do Extension, *TRB,* p. 10.

Practice Options/Homework Suggestions

- Internet (go.hrw.com, Keyword: WN3 ESFUERZOS-LEC)
- Have students read **"En la noche,"** pp. 22–29.
- Have students study **Vocabulario esencial, "En la noche,"** p. 73. ◆ ●

▲ = Advanced Learners ◆ = Slower Pace Learners ● = Special Learning Needs ■ = Heritage Speakers

Teacher's Name ______________________ Class ______________________ Date ____________

COLECCIÓN 1

Esfuerzos heroicos

DAY 10 50-MINUTE LESSON PLAN

STANDARDS FOR FOREIGN LANGUAGE LEARNING: DAY 10

Lectura

Communication 1.1: Students engage in conversations, provide and obtain information, express feelings and emotions, and exchange opinions.

Communication 1.2: Students understand and interpret written and spoken language on a variety of topics.

Connections 3.1: Students reinforce and further their knowledge of other disciplines through the foreign language.

Comparisons 4.1: Students demonstrate understanding of the nature of language through comparisons of the language studied and their own.

CORE INSTRUCTION

Warm-Up

- (5 min.) Have students read **Antes de leer,** p. 21.

Lectura

Teach

- (10 min.) Have students do **Punto de partida,** *TRB,* p. 11.
- (30 min.) Have students begin reading **"En la noche,"** pp. 22–29. See Techniques for Handling the Reading, *TRB,* p. 12.

Wrap-Up

- (5 min.) Have students use the expressions in **Así se dice,** p. 30, to talk about the story. See **Así se dice,** *TRB,* p. 12.

OPTIONAL RESOURCES

- (10 min.) Read aloud **"En la noche,"** Summary, *TRB,* p. 11. ◆ ●
- (20 min.) Have students do **Para hispanohablantes,** *TRB,* p. 13. ■
- (20 min.) Have students do **Para angloparlantes,** *TRB,* p. 13.

Practice Options/Homework Suggestions

- Internet (go.hrw.com, Keyword: WN3 ESFUERZOS-LEC)
- *Cuaderno de práctica,* Activities 1–4, pp. 3–4
- Have students finish reading **"En la noche,"** pp. 22–29.

▲ = Advanced Learners ◆ = Slower Pace Learners ● = Special Learning Needs ■ = Heritage Speakers

Esfuerzos heroicos

DAY 11 50-MINUTE LESSON PLAN

STANDARDS FOR FOREIGN LANGUAGE LEARNING: DAY 11

Lectura

Communication 1.2: Students understand and interpret written and spoken language on a variety of topics.

Connections 3.1: Students reinforce and further their knowledge of other disciplines through the foreign language.

CORE INSTRUCTION

Warm-Up

- (5 min.) Have students read **Conoce al escritor,** p. 29.

Lectura

Teach

- (10 min.) Have students do **Crea significados: Primeras impresiones,** p. 30.
- (20 min.) Have pairs of students do **Interpretaciones del texto,** p. 30.
- (10 min.) Have students do **Conexiones con el texto,** p. 30.

Wrap-Up

- (5 min.) Do **Repaso del texto,** p. 30.

OPTIONAL RESOURCES

- (20 min.) Have students practice the Reading Strategy, *TRB,* pp. 236–237. ◆ ●

Practice Options/Homework Suggestions

- Internet (go.hrw.com, Keyword: WN3 ESFUERZOS-LEC)
- Have students study **Vocabulario adicional, "En la noche,"** *TRB,* p. 297. ◆ ●

▲ = Advanced Learners ◆ = Slower Pace Learners ● = Special Learning Needs ■ = Heritage Speakers

Teacher's Name ________________ Class ________________ Date ________

COLECCIÓN 1

Esfuerzos heroicos

DAY 12 50-MINUTE LESSON PLAN

STANDARDS FOR FOREIGN LANGUAGE LEARNING: DAY 12

Lectura

Communication 1.2: Students understand and interpret written and spoken language on a variety of topics.

Communities 5.2: Students show evidence of becoming life-long learners by using the language for personal enjoyment and enrichment.

CORE INSTRUCTION

Warm-Up

- (5 min.) Have students do **Vocabulario en contexto,** Activity B, pp. 47–48.

Lectura

Teach

- (5 min.) Go over answers for Activity B, pp. 47–48.
- (15 min.) Have students do Thinking Critically, *TRB,* p. 23.
- (10 min.) Have students do Activity C, p. 48. See Activity C, *TRB,* p. 23.
- (10 min.) Have students do the first Additional Practice, *TRB,* p. 23.

Wrap-Up

- (5 min.) Have students do the second Additional Practice, *TRB,* p. 23.

OPTIONAL RESOURCES

- (25 min.) Have students do **Relato de aventuras,** p. 31. ▲
- (30 min.) Have groups of students do **Representación artística,** p. 31.

Practice Options/Homework Suggestions

- Internet (go.hrw.com, Keyword: WN3 ESFUERZOS-LEC)
- Have students do Additional Practice, *TRB,* p. 10. ◆ ●
- Have students do Extension, *TRB,* p. 10. ▲
- Have students study for **Prueba de lectura.**

▲ = Advanced Learners ◆ = Slower Pace Learners ● = Special Learning Needs ■ = Heritage Speakers

Teacher's Name ______________ Class ______________ Date ______________

COLECCIÓN 1

Esfuerzos heroicos

DAY 13 50-MINUTE LESSON PLAN

STANDARDS FOR FOREIGN LANGUAGE LEARNING: DAY 13

Lectura

Communication 1.2: Students understand and interpret written and spoken language on a variety of topics.

Connections 3.1: Students reinforce and further their knowledge of other disciplines through the foreign language.

CORE INSTRUCTION

Warm-Up

- (5 min.) Have students review **"En la noche,"** pp. 22–29.

Lectura
Assess

- (30 min.) Give **Prueba de lectura: "En la noche,"** *Assessment Program,* pp. 5–6.

Lectura
Teach

- (10 min.) Have students do **Punto de partida,** p. 32.

Wrap-Up

- (5 min.) Review **Elementos de literatura,** p. 32.

OPTIONAL RESOURCES

- (15 min.) Have students do **Toma nota,** p. 32.
- (15 min.) Have students read **"Trabajo de campo,"** pp. 33–37.
- (10 min.) Have students read **Conoce a la escritora,** p. 37.
- (10 min.) Have students listen to the audio of **"Trabajo de campo,"** on Audio CD 3, Track 1.
- (10 min.) Have students do **Estrategias para leer**, *TRB,* p. 14. ◆ ●
- (10 min.) Have students do Techniques for Handling the Reading, *TRB,* pp. 14–15. ◆ ●

Practice Options/Homework Suggestions

- Internet (go.hrw.com, Keyword: WN3 ESFUERZOS-LEC)
- Have students read **"Trabajo de campo,"** pp. 33–37. See **"Trabajo de campo,"** *TRB,* pp. 14–16.
- Have students do **Para hispanohablantes,** *TRB,* p. 16. ■
- Have students practice the Reading Strategy, pp. 238–239. ◆ ●

▲ = Advanced Learners ◆ = Slower Pace Learners ● = Special Learning Needs ■ = Heritage Speakers

Teacher's Name ______________ Class ______________ Date ______________

COLECCIÓN 1

Esfuerzos heroicos

DAY 14 50-MINUTE LESSON PLAN

STANDARDS FOR FOREIGN LANGUAGE LEARNING: DAY 14

Lectura/Vocabulario

Communication 1.2: Students understand and interpret written and spoken language on a variety of topics.

Communication 1.3: Students present information, concepts, and ideas to an audience of listeners or readers on a variety of topics.

Comparisons 4.1: Students demonstrate understanding of the nature of language through comparisons of the language studied and their own.

Communities 5.1: Students use the language both within and beyond the school setting.

CORE INSTRUCTION

Warm-Up

- (5 min.) Have students review the expressions in **Así se dice,** p. 38.

Lectura/Vocabulario

Teach

- (5 min.) Have students answer **Primeras impresiones,** p. 38. See **Primeras impresiones,** *TRB,* p. 15.
- (10 min.) Have students do **Interpretaciones del texto,** p. 38. See **Interpretaciones del texto,** *TRB,* p. 15.
- (10 min.) Have students answer **Conexiones con el texto** and **Preguntas al texto,** p. 38. See **Conexiones con el texto** and **Preguntas al texto,** *TRB,* p. 15.
- (5 min.) Have students do **Vocabulario,** Activity D, p. 48. See Activity D, *TRB,* pp. 23–24.
- (10 min.) Have students do Activity E, p. 49. You may wish to use Audio CD 1, Track 4. See script, *TRB,* p. 24.

Wrap-Up

- (5 min.) Do Group Work, *TRB,* p. 24.

OPTIONAL RESOURCES

- (5 min.) Read aloud **"Trabajo de campo,"** Summary, *TRB,* p. 14. ◆ ●
- (15 min.) Have students do **Para angloparlantes,** *TRB,* p. 16.
- (30 min.) Have students do **Opciones: Prepara tu portafolio,** p. 39. See *TRB,* pp. 15–16.

Practice Options/Homework Suggestions

- Internet (go.hrw.com, Keyword: WN3 ESFUERZOS-LEC)
- *Cuaderno de práctica,* Activities, 1–3, pp. 5–6
- Have students answer the questions in **Repaso del texto,** p. 38.
- Have students study **Vocabulario esencial, "Trabajo de campo,"** p. 73. ◆ ●
- Have students study **Vocabulario adicional, "Trabajo de campo,"** *TRB,* p. 297. ◆ ●
- Have students study for **Prueba de lectura.**

▲ = Advanced Learners ◆ = Slower Pace Learners ● = Special Learning Needs ■ = Heritage Speakers

Esfuerzos heroicos

DAY 15 50-MINUTE LESSON PLAN

STANDARDS FOR FOREIGN LANGUAGE LEARNING: DAY 15

Lectura/Vocabulario

Communication 1.2: Students understand and interpret written and spoken language on a variety of topics.

Connections 3.1: Students reinforce and further their knowledge of other disciplines through the foreign language.

CORE INSTRUCTION

Warm-Up

- (5 min.) Have students review **"Trabajo de campo,"** pp. 33–37.

Lectura

Assess

- (20 min.) *Assessment Program:* **Prueba de lectura: "Trabajo de campo,"** pp. 7–8.

Vocabulario

Teach

- (10 min.) Present **Mejora tu vocabulario: Los sinónimos,** p. 49. See **Mejora tu vocabulario: Los sinónimos,** *TRB,* p. 24.
- (10 min.) Have students do Activity F, p. 49. See Activity F, *TRB,* pp. 24–25.

Wrap-Up

- (5 min.) Have students do Challenge, *TRB,* p. 24.

OPTIONAL RESOURCES

- (10 min.) Have students do **Para angloparlantes,** *TRB,* p. 25.
- (15 min.) Have students do Thinking Critically, *TRB,* p. 25. ▲

Practice Options/Homework Suggestions

- Have students read **Tono y registro,** p. 50.
- Have students study **Vocabulario esencial, Mejora tu vocabulario,** p. 73. ◆ ●
- *Cuaderno de práctica,* Activities 1–3, pp. 10–11

▲ = Advanced Learners ◆ = Slower Pace Learners ● = Special Learning Needs ■ = Heritage Speakers

Teacher's Name ______________________ Class ____________________ Date __________

Esfuerzos heroicos

DAY 16 50-MINUTE LESSON PLAN

STANDARDS FOR FOREIGN LANGUAGE LEARNING: DAY 16

Vocabulario/Gramática

Communication 1.2: Students understand and interpret written and spoken language on a variety of topics.

Communication 1.3: Students present information, concepts, and ideas to an audience of listeners or readers on a variety of topics.

Communities 5.1: Students use the language both within and beyond the school setting.

CORE INSTRUCTION

Warm-Up

- (5 min.) Have students discuss **Tono y registro,** p. 50.

Vocabulario

Teach

- (5 min.) Present **Tono y registro,** *TRB,* p. 25.
- (20 min.) Have students do Activities G, H, and I , p. 51.

Gramática

Teach

- (10 min.) Present **Los pronombres de complemento preposicional,** p. 56. See **Los pronombres de complemento preposicional,** *TRB,* p. 27.
- (5 min.) Have students do Activity G, p. 56.

Wrap-Up

- (5 min.) Have students review **¿Te acuerdas?,** p. 56. Ask students to think of sentences using **conmigo, contigo, para mí,** and **a ti.**

OPTIONAL RESOURCES

- (20 min.) Have students do **Para hispanohablantes,** *TRB,* p. 25. ■

Practice Options/Homework Suggestions

- *Cuaderno de práctica,* Activities 8–10, p. 15
- *Cuaderno de práctica,* **Vocabulario adicional,** p.133
- Have students study for **Prueba de vocabulario.**

▲ = Advanced Learners ◆ = Slower Pace Learners ● = Special Learning Needs ■ = Heritage Speakers

Esfuerzos heroicos

DAY 17 50-MINUTE LESSON PLAN

STANDARDS FOR FOREIGN LANGUAGE LEARNING: DAY 17

Vocabulario/Gramática

Communication 1.2: Students understand and interpret written and spoken language on a variety of topics.

Comparisons 4.1: Students demonstrate understanding of the nature of language through comparisons of the language studied and their own.

CORE INSTRUCTION

Warm-Up

- (5 min.) Have students review **Vocabulario,** pp. 47–51.

Vocabulario

Assess

- (20 min.) Give **Prueba de vocabulario,** *Assessment Program,* pp. 11–12.

Gramática

Teach

- (5 min.) Present **Los pronombres reflexivos,** pp. 56–57. See **Los pronombres reflexivos,** *TRB,* p. 27.
- (15 min.) Have students do Activities H and I, p. 57.

Wrap-Up

- (5 min.) In pairs, have students do Additional Practice, *TRB,* p. 28.

OPTIONAL RESOURCES

- (20 min.) Have students do **Ampliación, Hoja de práctica 1-D, Más sobre los pronombres reflexivos,** *TRB,* p. 280. ◆ ●
- (20 min.) Have students do **Para hispanohablantes,** *TRB,* p. 29. ■

Practice Options/Homework Suggestions

- *Cuaderno de práctica,* **Ampliación, Hoja de práctica 1-D,** p. 137 ◆ ●
- *Cuaderno de práctica,* Activities 11–12, p. 16
- Have students study for **Prueba de gramática.**

Assessment Options

- *Assessment Program:* **Prueba de vocabulario adicional,** p. 249

▲ = Advanced Learners ◆ = Slower Pace Learners ● = Special Learning Needs ■ = Heritage Speakers

Esfuerzos heroicos

DAY 18 50-MINUTE LESSON PLAN

STANDARDS FOR FOREIGN LANGUAGE LEARNING: DAY 18

Gramática

Communication 1.2: Students understand and interpret written and spoken language on a variety of topics.

Comparisons 4.1: Sutdents demonstrate an understanding of the nature of language through comparisons of the language studied and their own.

CORE INSTRUCTION

Warm-Up

- (5 min.) Have students review **Gramática,** pp. 56–57.

Gramática

Assess

- (15 min.) Give **Prueba de gramática: Los pronombres de complemento y los pronombres relexivos,** *Assessment Program,* p. 14.

Gramática

Teach

- (10 min.) Present **Los pronombres posesivos,** p. 58. See **Los pronombres posesivos,** *TRB,* p. 28.
- (5 min.) Have students do Activity J, p. 58.
- (5 min.) Present **Los pronombres demostrativos,** p. 59. See **Los pronombres demostrativos,** *TRB,* p. 28.
- (5 min.) Have students do Activity K, p. 59.

Wrap-Up

- (5 min.) Have students do Activity L, p. 60.

OPTIONAL RESOURCES

- (10 min.) Have students do Activity M, p. 60.
- (30 min.) Have students read **Comparación y contraste,** pp. 60–61, and do Activities A, B, and C, p. 61.

Practice Options/Homework Suggestions

- *Cuaderno de práctica,* Activities 13–16, pp. 17–18
- Have students study for **Prueba de gramática.**

Assessment Options

- *Assessment Program:* **Prueba de gramática 1-D, Más sobre los pronombres reflexivos,** p. 253

▲ = Advanced Learners ◆ = Slower Pace Learners ● = Special Learning Needs ■ = Heritage Speakers

Teacher's Name ______________________ Class ______________________ Date __________

COLECCIÓN

Esfuerzos heroicos

DAY 19 50-MINUTE LESSON PLAN

STANDARDS FOR FOREIGN LANGUAGE LEARNING: DAY 19

Gramática

Comparisons 4.1: Students demonstrate understanding of the nature of language through comparisons of the language studied and their own.

CORE INSTRUCTION

Warm-Up

- (5 min.) Have students review **Gramática,** pp. 58–59.

Gramática

Assess

- (15 min.) Give **Prueba de gramática: Los pronombres posesivos y los pronombres demostrativos,** *Assessment Program,* p. 15.

Gramática

Teach

- (25 min.) Present **Comparación y contraste,** pp. 60–61, and have students do Activities A and B, p. 61.

Wrap-Up

- (5 min.) As a class, create two sentences for Activity C, p. 61.

OPTIONAL RESOURCES

- (10 min.) Have students discuss the introduction of **Panorama cultural,** p. 40. See **Panorama cultural,** *TRB,* p. 17.
- (5 min.) Play Audio CD 1, Tracks 1–3, and have students listen to the interviews.

Practice Options/Homework Suggestions

- Have students read **Panorama cultural,** pp. 40–41.
- Have students do **Para pensar y hablar,** p. 41.
- Have students do Activity C, p. 61.
- *Cuaderno de práctica,* Activities 17–18, p. 19
- Have students study for **Prueba de comparación y contraste.**

▲ = Advanced Learners ◆ = Slower Pace Learners ● = Special Learning Needs ■ = Heritage Speakers

Teacher's Name ____________ Class ____________ Date ________

COLECCIÓN 1

Esfuerzos heroicos

DAY 20 50-MINUTE LESSON PLAN

STANDARDS FOR FOREIGN LANGUAGE LEARNING: DAY 20

Gramática/Panorama cultural

Communication 1.2: Students understand and interpret written and spoken language on a variety of topics.

Comparisons 4.2: Students demonstrate understanding of the concept of culture through comparisons of the cultures studied and their own.

CORE INSTRUCTION

Warm-Up

- (5 min.) Have students review **Comparación y contraste,** pp. 60–61.

Gramática

Assess

- (20 min.) Give **Prueba de comparación y contraste,** *Assessment Program,* p. 16.

Panorama cultural

Teach

- (15 min.) Show **Panorama cultural,** *Video Program* (Videocassette 1). See **Panorama cultural,** Teaching Suggestions, **Mientras lo ves,** *Video Guide,* p. 4.
- (5 min.) Have students do **Hoja de actividades 3, Mientras lo ves,** *Video Guide,* p. 8.

Wrap-Up

- (5 min.) Have students answer the questions in **Hoja de actividades 3, Después de ver,** *Video Guide,* p. 8.

OPTIONAL RESOURCES

- (20 min.) Have students do **Panorama cultural,** Teaching Suggestions, **Antes de ver,** *Video Guide,* p. 4.
- (20 min.) Have students do **Panorama cultural,** Teaching Suggestions, **Después de ver,** *Video Guide,* p. 4.
- (10 min.) Have students do **Panorama cultural, Hoja de actividades 3, Antes de ver,** *Video Guide,* p. 8.

Practice Options/Homework Suggestions

- Have students do Culture Link, *TRB,* p. 17.
- Have students do Community Link, *TRB,* p. 17.

▲ = Advanced Learners ◆ = Slower Pace Learners ● = Special Learning Needs ■ = Heritage Speakers

Teacher's Name ______________________ Class ____________________ Date __________

COLECCIÓN 1

Esfuerzos heroicos

DAY 21 50-MINUTE LESSON PLAN

STANDARDS FOR FOREIGN LANGUAGE LEARNING: DAY 21

Comunidad y oficio

Communication 1.1: Students engage in conversations, provide and obtain information, express feelings and emotions, and exchange opinions.

Connections 3.1: Students reinforce and further their knowledge of other disciplines through the foreign language.

Communities 5.2: Students show evidence of becoming life-long learners by using the language for personal enjoyment and enrichment.

CORE INSTRUCTION

Warm-Up

- (5 min.) Have students answer the questions in Getting Started, *TRB,* p. 22.

Comunidad y oficio

Teach

- (10 min.) Have students read **Comunidad y oficio,** p. 46. See History Link, *TRB,* p. 22.
- (10 min.) Have students do **Comunidad y oficio,** Teaching Suggestions, **Antes de ver,** *Video Guide,* p. 5.
- (10 min.) Show **Comunidad y oficio,** *Video Program* (Videocassette 1). See **Comunidad y oficio,** Teaching Suggestions, **Mientras lo ves,** *Video Guide,* p. 5.
- (10 min.) Have students do **Hoja de actividades 4, Mientras lo ves,** *Video Guide,* p. 9.

Wrap-Up

- (5 min.) Have students answer the questions in **Hoja de actividades 4, Después de ver,** *Video Guide,* p. 9.

OPTIONAL RESOURCES

- (30 min.) Have students do the first Thinking Critically, *TRB,* p. 22. ▲
- (15 min.) Have students do the second Thinking Critically, *TRB,* p. 22. ▲
- (30 min.) Have students do Group work, *TRB,* p. 22.
- (5 min.) Have students do Pair Work, *TRB,* p. 22.

Practice Options/Homework Suggestions

- Internet (go.hrw.com, Keyword: WN3 ESFUERZOS-CYO)
- Have students do **Investigaciones,** Activity A or B, p. 46.
- Have students do **Comunidad y oficio,** Teaching Suggestions, **Después de ver,** *Video Guide,* p. 5.

▲ = Advanced Learners ◆ = Slower Pace Learners ● = Special Learning Needs ■ = Heritage Speakers

Teacher's Name ______________ Class ______________ Date ______________

COLECCIÓN

Esfuerzos heroicos

DAY 22 50-MINUTE LESSON PLAN

STANDARDS FOR FOREIGN LANGUAGE LEARNING: DAY 22

Ortografía

Comparisons 4.1: Students demonstrate understanding of the nature of language through comparisons of the language studied and their own.

CORE INSTRUCTION

Warm-Up

- (5 min.) Have students read about **El uso de las mayúsculas** in **Letra y sonido,** p. 62.

Ortografía

Teach

- (10 min.) Present **El uso de las mayúsculas,** p. 62. See **El uso de las mayúsculas,** *TRB,* p. 30.
- (15 min.) Have students do Activities A and B, p. 63.
- (10 min.) Present **La acentuación,** p. 63. See **La acentuación,** *TRB,* p. 31.
- (5 min.) Have students do Activity C, pp. 64–65.

Wrap-Up

- (5 min.) Review **El acento diacrítico** by reading some of the example sentences on pp. 63–64 in random order and having the students raise their hand when they hear a targeted word that has a written accent mark.

OPTIONAL RESOURCES

- (10 min.) Have students do Additional Practice, *TRB,* p. 30. ▲
- (5 min.) Have students do Activity D, p. 65.
- (10 min.) Have students do **Ampliación, Hoja de práctica 1-E, La sinalefa y el enlace,** *TRB,* p. 281. ◆ ●

Practice Options/Homework Suggestions

- *Cuaderno de práctica,* Activities 1–5, pp. 20–21
- *Cuaderno de práctica,* **Ampliación, Hoja de práctica 1-E,** p. 138 ◆ ●
- Have students do Activity D, p. 65.
- Have students study **La diéresis,** p. 65.
- Have students study for **Prueba de ortografía.**

▲ = Advanced Learners ◆ = Slower Pace Learners ● = Special Learning Needs ■ = Heritage Speakers

Teacher's Name ______________ Class ______________ Date ______________

Esfuerzos heroicos

DAY 23 50-MINUTE LESSON PLAN

STANDARDS FOR FOREIGN LANGUAGE LEARNING: DAY 23

Ortografía

Communication 1.2: Students understand and interpret written and spoken language on a variety of topics.

Comparisons 4.1: Students demonstrate understanding of the nature of language through comparisons of the language studied and their own.

CORE INSTRUCTION

Warm-Up

- (5 min.) Have students review **El uso de las mayúsculas** and **El acento diacrítico,** pp. 62–64.

Ortografía
Teach

- (5 min.) Present **La diéresis,** p. 65. See **La diéresis,** *TRB,* p. 31.
- (5 min.) Have students do Activity E, p. 65.
- (10 min.) Give **Dictado,** Activity B, p. 65. You may wish to use Audio CD 1, Track 6. See script, *TRB,* pp. 31–32.
- (10 min.) Review **Ortografía,** pp. 62–65.

Ortografía
Assess

- (10 min.) Give **Prueba de ortografía,** *Assessment Program,* p. 18.

Wrap-Up

- (5 min.) Review the chart on **El acento diacrítico,** pp. 63–64. Ask students to think of other examples.

OPTIONAL RESOURCES

- (20 min.) Have students do **Para hispanohablantes,** *TRB,* p. 32. ■
- (20 min.) Have students do **Para angloparlantes,** *TRB,* p. 32.
- (15 min.) Give **Dictado,** Activity A, p. 65. You may use Audio CD 1, Track 5. See *TRB,* pp. 31–32.

Practice Options/Homework Suggestions

- Have students read **Antes de escribir, Cuaderno del escritor,** p. 66.

Assessment Options

- *Assessment Program,* **Prueba de ortografía 1-E: La sinalefa y el enlace,** p. 254

▲ = Advanced Learners ◆ = Slower Pace Learners ● = Special Learning Needs ■ = Heritage Speakers

Teacher's Name ______________ Class ______________ Date ______________

COLECCIÓN 1

Esfuerzos heroicos

DAY 24 50-MINUTE LESSON PLAN

STANDARDS FOR FOREIGN LANGUAGE LEARNING: DAY 24

Taller del escritor

Communication 1.2: Students understand and interpret written and spoken language on a variety of topics.

Communication 1.3: Students present information, concepts, and ideas to an audience of listeners or readers on a variety of topics.

Connections 3.1: Students reinforce and further their knowledge of other disciplines through the foreign language.

Communities 5.1: Students use the language both within and beyond the school setting.

CORE INSTRUCTION

Warm-Up

- (5 min.) Have students review their work on the portfolio suggestions after each reading selection. See **Opciones: prepara tu portafolio,** pp. 13, 31, and 39.

Taller del escritor

Teach

- (10 min.) Introduce **Taller del escritor,** p. 66. Have students read **La narración: Episodio autobiográfico,** p. 66.
- (10 min.) See **Episodio autobiográfico,** Presenting the Workshop, *TRB,* p. 33.
- (10 min.) Introduce **Antes de escribir,** items 1–3, pp. 66–67. See **Antes de escribir,** *TRB,* p. 33.
- (10 min.) Have students do **Pautas para temas,** pp. 66.

Wrap-Up

- (5 min.) Have students volunteers read samples of their **Escritura libre,** p. 66.

OPTIONAL RESOURCES

- (20 min.) Have students read **A leer por tu cuenta: "Soneto 149,"** pp. 44–45. See *TRB,* pp. 20–21. You may wish to use Audio CD 3, Track 2. ▲
- (5 min.) Show **Lectura: "Soneto 149,"** *Video Program* (Videocassette 1). See Teaching Suggestions, **Lectura,** *Video Guide,* p. 2. ▲
- (20 min.) Have students compare the authors in this collection and their works, pp. 10, 29, and 37. ▲

Practice Options/Homework Suggestions

- Have students choose the theme of their autobiographical event, p. 66. See **Antes de escribir,** *TRB,* p. 33.
- Have students do **Hoja de actividades 1, "Soneto 149,"** *Video Guide,* p. 6. ▲

▲ = Advanced Learners ◆ = Slower Pace Learners ● = Special Learning Needs ■ = Heritage Speakers

Esfuerzos heroicos

DAY 25 50-MINUTE LESSON PLAN

STANDARDS FOR FOREIGN LANGUAGE LEARNING: DAY 25

Taller del escritor

Communication 1.3: Students present information, concepts, and ideas to an audience of listeners or readers on a variety of topics.

Connections 3.1: Students reinforce and further their knowledge of other disciplines through the foreign language.

CORE INSTRUCTION

Warm-Up

- (5 min.) Have students review and revise their writing for **Escritura libre,** p. 66.

Taller del escritor

Teach

- (10 min.) Discuss **Objetivo y público,** p. 67.
- (15 min.) Have students do **Compilación de datos,** p. 67.
- (15 min.) Present **El borrador: Organización,** p. 68. See **El borrador,** *TRB,* p. 33.

Wrap-Up

- (5 min.) Have student volunteers write their outlines on the chalkboard.

OPTIONAL RESOURCES

- (20 min.) Have students read **Elementos de literatura,** p. 42. See *TRB,* pp. 18–19.

Practice Options/Homework Suggestions

- Have students edit their own writing for **Escritura libre.**
- Have students make an outline for their writing.

Assessment Options

- *Assessment Program,* **Prueba de lectura: "Soneto 149,"** pp. 9–10

▲ = Advanced Learners ◆ = Slower Pace Learners ● = Special Learning Needs ■ = Heritage Speakers

Teacher's Name ______________________ Class ______________________ Date ______________

COLECCIÓN

Esfuerzos heroicos

DAY 26 50-MINUTE LESSON PLAN

STANDARDS FOR FOREIGN LANGUAGE LEARNING: DAY 26

Taller del escritor

Communication 1.3: Students present information, concepts, and ideas to an audience of listeners or readers on a variety of topics.

Connections 3.1: Students reinforce and further their knowledge of other disciplines through the foreign language.

Communities 5.1: Students use the language both within and beyond the school setting.

CORE INSTRUCTION

Warm-Up

- (5 min.) Have students review and revise their outlines, p. 68.

Taller del escritor

Teach

- (5 min.) Present **Relaciona ideas,** p. 68, and the expressions in **Así se dice,** p. 69.
- (30 min.) Have students do **El desarrollo,** p. 68. Have students work on writing descriptions and dialogs.

Wrap-Up

- (10 min.) Have students use the expressions in **Así se dice,** p. 69, to evaluate their peers' writing.

OPTIONAL RESOURCES

- (10 min.) Have students use **Pautas para redactar,** p. 68.

Practice Options/Homework Suggestion

- Have students work on their drafts.

▲ = Advanced Learners ◆ = Slower Pace Learners ● = Special Learning Needs ■ = Heritage Speakers

Esfuerzos heroicos

DAY 27 50-MINUTE LESSON PLAN

STANDARDS FOR FOREIGN LANGUAGE LEARNING: DAY 27

Taller del escritor

Communication 1.2: Students understand and interpret written and spoken language on a variety of topics.

Communication 1.3: Students present information, concepts, and ideas to an audience of listeners or readers on a variety of topics.

Communities 5.1: Students use the language both within and beyond the school setting.

CORE INSTRUCTION

Warm-Up

- (5 min.) Have students discuss **Pautas de evaluación,** p. 69.

Taller del escritor

Teach

- (10 min.) Using the guidelines in **Pautas de evaluación** and **Técnicas de revisión,** have students revise their writing.
- (10 min.) Have students compare the **Modelos,** pp. 69–70.
- (15 min.) Have students do **Corrección de pruebas,** p. 70. See **Corrección de pruebas,** *TRB,* p. 34.

Wrap-Up

- (10 min.). In pairs, have students do **Reflexión,** p. 70, using the expressions in **Así se dice,** p. 70.

OPTIONAL RESOURCES

- (15 min.) Have students do Additional Practice, *TRB,* p. 34. ▲
- (25 min.) Have students do Pair Work, *TRB,* p. 34. ◆ ●
- (20 min.) Have students do **Publicación,** p. 70. See **Publicación,** *TRB,* p. 34. ▲
- (5 min.) Have students do Closure, *TRB,* p. 34.

Practice Options/Homework Suggestions

- Have students do **A ver si puedo...,** pp. 71–72.

▲ = Advanced Learners ◆ = Slower Pace Learners ● = Special Learning Needs ■ = Heritage Speakers

Esfuerzos heroicos

DAY 28 50-MINUTE LESSON PLAN

STANDARDS FOR FOREIGN LANGUAGE LEARNING: DAY 28

A ver si puedo...
Communication 1.2: Students understand and interpret written and spoken language on a variety of topics.

Connections 3.1: Students reinforce and further their knowledge of other disciplines through the foreign language.

CORE INSTRUCTION

Warm-Up

- (5 min.) Have students review the objectives listed on the Collection Opener, p. xxii.

A ver si puedo...

Review

- (10 min.) Have students do **Lectura,** Activities A and B, p. 71.
- (10 min.) Have students do **Cultura,** Activity C, p. 71.
- (10 min.) Have students do **Comunicación,** Activities D, E, F, and G, pp. 71–72.
- (10 min.) Have students do **Escritura,** Activities H and I, p. 72.

Wrap-Up

- (5 min.) Answer any questions about either of the two chapter exams.

OPTIONAL RESOURCES

- (50 min.) Have students read **Enlaces literarios: La prosa didáctica medieval,** pp. 74–78. See *TRB,* pp. 37–39. ▲
- (20 min.) Have students do **Comprensión del texto,** p. 79. See **Comprensión del texto,** *TRB,* p. 40. ▲
- (25 min.) Have students do **Análisis del texto,** p. 79. See **Análisis del texto,** *TRB,* p. 40. ▲
- (25 min.) Have students do **Más allá del texto,** p. 79. See **Más allá del texto,** *TRB,* p. 40. ▲

Practice Options/Homework Suggestions

- Have students study for the **Examen de lengua.**

▲ = Advanced Learners ◆ = Slower Pace Learners ● = Special Learning Needs ■ = Heritage Speakers

Esfuerzos heroicos

DAY 29 50-MINUTE LESSON PLAN

CORE INSTRUCTION

Assess

- (50 min.) Give **Colección 1 Examen de lengua,** *Assessment Program,* pp. 25–32.

OPTIONAL RESOURCES

- (50 min.) Give **Examen de lectura: de *Autobiografía de un esclavo,* "En la noche" y "Trabajo de campo,"** *Assessment Program,* pp. 19–24. To allow students more time to take the exam, either **Examen** may be given over two class periods.

Practice Options/Homework Suggestions

- Have students study for the **Examen de lectura: de *Autobiografía de un esclavo,*" "En la noche" y "Trabajo de campo."**

Assessment Options

- *Assessment Program,* Performance Assessment, p. 295

▲ = Advanced Learners ◆ = Slower Pace Learners ● = Special Learning Needs ■ = Heritage Speakers

Esfuerzos heroicos

DAY 30 50-MINUTE LESSON PLAN

CORE INSTRUCTION

Assess

- (50 min.) Give **Examen de lectura: de *Autobiografía de un esclavo,* "En la noche" y "Trabajo de campo,"** *Assessment Program,* pp. 19–24.

OPTIONAL RESOURCES

- (50 min.) Give **Colección 1 Examen de lengua,** *Assessment Program,* pp. 25–32. To allow students more time to take the exam, either **Examen** may be given over two class periods.

Practice Options/Homework Suggestions

- Internet (go.hrw.com, Keyword: WN3 ESFUERZOS)

Assessment Options

- *Assessment Program,* Performance Assessment, p. 295

▲ = Advanced Learners ◆ = Slower Pace Learners ● = Special Learning Needs ■ = Heritage Speakers

Teacher's Name ______________________ Class ______________________ Date ______________

COLECCIÓN **2**

Lazos de amistad

DAY 1 50-MINUTE LESSON PLAN

STANDARDS FOR FOREIGN LANGUAGE LEARNING: DAY 1

Lectura

Communication 1.1: Students engage in conversations, provide and obtain information, express feelings and emotions, and exchange opinions.

Communication 1.2: Students understand and interpret written and spoken language on a variety of topics.

Communication 1.3: Students present information, concepts, and ideas to an audience of listeners or readers on a variety of topics.

Connections 3.1: Students reinforce and further their knowledge of other disciplines through the foreign language.

Communities 5.1: Students use the language both within and beyond the school setting.

Communities 5.2: Students show evidence of becoming life-long learners by using the language for personal enjoyment and enrichment.

CORE INSTRUCTION

Warm-Up

- (5 min.) Have students read the objectives on the Collection Opener, p. 80. See Collection Overview, *TRB,* p. 44.

Lectura

Teach

- (10 min.) Have students discuss **Punto de partida,** p. 82. See **Punto de partida,** *TRB,* p. 45.
- (10 min.) Have students do **Comparte tus ideas,** p. 82. See **Comparte tus ideas,** *TRB,* p. 45.
- (5 min.) Discuss **Elementos de literatura,** p. 82. See **Elementos de literatura,** *TRB,* p. 45.
- (15 min.) Have students begin reading **"Cadena rota"** and **"Naranjas,"** pp. 83–91.

Wrap-Up

- (5 min.) Do **Crea significados: Primeras impresiones,** p. 92.

OPTIONAL RESOURCES

- (10 min.) See Presentation Suggestions, item one, *TRB,* p. 44.
- (10 min.) See Presentation Suggestions, item two, *TRB,* p. 44.
- (10 min.) See Presentation Suggestions, item three, *TRB,* p. 44.
- (10 min.) Have students discuss the information in **Antes de leer,** *TRB,* p. 45.
- (5 min.) See Techniques for Handling the Reading, *TRB,* pp. 45–46. ◆ ●
- (10 min.) Read aloud **"Cadena rota"** and **"Naranjas,"** Summary, *TRB,* p. 45. ◆ ●

Practice Options/Homework Suggestions

- Internet (go.hrw.com, Keyword: WN3 AMISTAD-LEC)
- Have students read **"Cadena rota"** and **"Naranjas,"** pp. 83–91.
- Have students practice the Reading Strategy, *TRB,* pp. 243–245 and pp. 246–248. ◆ ●
- Have students study **Vocabulario esencial, "Cadena rota"** and **"Naranjas,"** *TRB,* p. 139. ◆ ●

▲ = Advanced Learners ◆ = Slower Pace Learners ● = Special Learning Needs ■ = Heritage Speakers

Teacher's Name ______________________ Class ______________________ Date ____________

COLECCIÓN 2

Lazos de amistad

DAY 2 50-MINUTE LESSON PLAN

STANDARDS FOR FOREIGN LANGUAGE LEARNING: DAY 2

Lectura/Vocabulario

Communication 1.1: Students engage in conversations, provide and obtain information, express feelings and emotions, and exchange opinions.

Communication 1.2: Students understand and interpret written and spoken language on a variety of topics.

Connections 3.1: Students reinforce and further their knowledge of other disciplines through the foreign language.

Comparisons 4.1: Students demonstrate understanding of the nature of language through comparisons of the language studied and their own.

CORE INSTRUCTION

Warm-Up

- (5 min.) Have students discuss **Una figura retórica,** p. 82, and **Glosario de términos literarios,** p. R9.

Lectura/Vocabulario

Teach

- (15 min.) Have students read aloud **"Cadena rota,"** pp. 83–90.
- (10 min.) Read **"Naranjas"** aloud to the students, p. 91. You may wish to use Audio CD 3, Track 3.
- (10 min.) Have students do **Vocabulario en contexto,** Activities A and B, p. 115. See **Vocabulario en contexto,** *TRB,* p. 60.

Wrap-Up

- (10 min.) Have students do the Pair Work activity, *TRB,* p. 60.

OPTIONAL RESOURCES

- (30 min.) Have students do Cooperative Learning, *TRB,* p. 60.
- (10 min.) Have pairs of students review the story in terms of **Diálogo con el texto,** p. 82. See **Diálogo con el texto,** *TRB,* p. 45.

Practice Options/Homework Suggestions

- Internet (go.hrw.com, Keyword: WN3 AMISTAD-LEC)
- Have students study **Vocabulario adicional, "Cadena rota"** and **"Naranjas,"** *TRB,* p. 298. ◆ ●
- *Cuaderno de práctica,* Activities 1–4, pp. 23–24
- *Advanced Placement Literature Preparation Book,* pp. 6–15 ▲

▲ = Advanced Learners ◆ = Slower Pace Learners ● = Special Learning Needs ■ = Heritage Speakers

Teacher's Name ______________________ Class ______________________ Date ____________

Lazos de amistad

DAY 3 50-MINUTE LESSON PLAN

STANDARDS FOR FOREIGN LANGUAGE LEARNING: DAY 3

Lectura

Communication 1.1: Students engage in conversations, provide and obtain information, express feelings and emotions, and exchange opinions.

Communication 1.2: Students understand and interpret written and spoken language on a variety of topics.

Connections 3.1: Students reinforce and further their knowledge of other disciplines through the foreign language.

Communities 5.1: Students use the language both within and beyond the school setting.

CORE INSTRUCTION

Warm-Up

- (5 min.) Have students read **Conoce al escritor,** p. 90.

Lectura

Teach

- (10 min.) Have students do **Crea significados: Interpretaciones del texto,** p. 92. See **Interpretaciones del texto,** *TRB,* p. 46.
- (15 min.) Have students describe Alfonso using the expressions in **Así se dice,** p. 92. See **Así se dice,** *TRB,* p. 46.
- (15 min.) Have students do **Cuaderno del escritor,** p. 93. See **Cuaderno del escritor,** *TRB,* pp. 46–47.

Wrap-Up

- (5 min.) Have students answer questions in **Conexiones con el texto,** p. 92.

OPTIONAL RESOURCES

- (20 min.) Have students do **Para hispanohablantes,** *TRB,* p. 47. ■
- (20 min.) Have students do **Para angloparlantes,** *TRB,* p. 47.

Practice Options/Homework Suggestions

- Internet (go.hrw.com, Keyword: WN3 AMISTAD-LEC)
- Have students do **Escribir/Resolver un problema,** p. 93. See **Escribir/Resolver un problema,** *TRB,* p. 47.
- *Cuaderno de práctica,* Activities 1–4, pp. 25–26
- *Cuaderno de práctica,* Activities 1–3, pp. 32–33, and Activity 5, p. 34
- Have students study for **Prueba de lectura.**

▲ = Advanced Learners ◆ = Slower Pace Learners ● = Special Learning Needs ■ = Heritage Speakers

Teacher's Name ______________________ Class ______________________ Date __________

COLECCIÓN 2

Lazos de amistad

DAY 4 50-MINUTE LESSON PLAN

STANDARDS FOR FOREIGN LANGUAGE LEARNING: DAY 4

Lectura

Communication 1.1: Students engage in conversations, provide and obtain information, express feelings and emotions, and exchange opinions.

Communication 1.2: Students understand and interpret written and spoken language on a variety of topics.

Connections 3.1: Students reinforce and further their knowledge of other disciplines through the foreign language.

Communities 5.1: Students use the language both within and beyond the school setting.

CORE INSTRUCTION

Warm-Up

- (5 min.) Have students review **Vocabulario esencial** for **"Cadena rota"** and **"Naranjas,"** p. 139.

Lectura

Assess

- (40 min.) Give **Prueba de lectura: "Cadena rota"** and **Prueba de lectura: "Naranjas,"** *Assessment Program,* pp. 41–42 and pp. 43–44.

Wrap-Up

- (5 min.) Have students do **Preguntas al texto,** p. 92.

OPTIONAL RESOURCES

- (30 min.) Have students do **Arte,** p. 93. See **Arte,** *TRB,* p. 47. Discuss drawings.
- (20 min.) Have students do **Dramatización,** p. 93. See **Dramatización,** *TRB,* p. 47.

Practice Options/Homework Suggestions

- Have students study **El adjetivo,** pp. 120–121.

▲ = Advanced Learners ◆ = Slower Pace Learners ● = Special Learning Needs ■ = Heritage Speakers

Teacher's Name ______________________ Class ______________________ Date ____________

COLECCIÓN 2

Lazos de amistad

DAY 5 50-MINUTE LESSON PLAN

STANDARDS FOR FOREIGN LANGUAGE LEARNING: DAY 5

Gramática

Communication 1.3: Students present information, concepts, and ideas to an audience of listeners or readers on a variety of topics.

Connections 3.1: Students reinforce and further their knowledge of other disciplines through the foreign language.

Comparisons 4.1: Students demonstrate understanding of the nature of language through comparisons of the language studied and their own.

CORE INSTRUCTION

Warm-Up

- (5 min.) Have students identify the adjectives used in **"Naranjas,"** p. 91.

Gramática

Teach

- (10 min.) Present **El adjetivo,** p. 120. See **El adjetivo,** *TRB,* p. 63.
- (10 min.) Present **Los adjetivos determinativos,** p. 121. See Language Note, *TRB,* p. 63.
- (15 min.) Have students do Activities A, B, and C, pp. 121–122. See Activities A, B, and C, *TRB,* p. 63.

Wrap-Up

- (10 min.) Have students do Additional Practice, *TRB,* p. 63.

OPTIONAL RESOURCES

- (20 min.) Have students do Additional Practice, *TRB,* p. 64. ◆ ●

Practice Options/Homework Suggestions

- Have students read **Cultura y lengua: Los mexicoamericanos,** pp. 94–95.
- *Cuaderno de práctica,* Activities 1–3, pp. 36–37
- Have students study for **Prueba de gramática.**

▲ = Advanced Learners ◆ = Slower Pace Learners ● = Special Learning Needs ■ = Heritage Speakers

Teacher's Name ______________________ Class ______________________ Date ______________

COLECCIÓN 2

Lazos de amistad

DAY 6 50-MINUTE LESSON PLAN

STANDARDS FOR FOREIGN LANGUAGE LEARNING: DAY 6

Gramática/Cultura y lengua

Communication 1.1: Students engage in conversations, provide and obtain information, express feelings and emotions, and exchange opinions.

Cultures 2.1: Students demonstrate an understanding of the relationship between the practices and perspectives of the culture studied.

Cultures 2.2: Students demonstrate an understanding of the relationship between the products and perspectives of the culture studied.

Connections 3.1: Students reinforce and further their knowledge of other disciplines through the foreign language.

Connections 3.2: Students acquire information and recognize the distinctive viewpoints that are only available through the foreign language and its cultures.

CORE INSTRUCTION

Warm-Up

- (5 min.) Have students review **Los adjetivos,** pp. 120–122.

Gramática

Assess

- (20 min.) Give **Prueba de gramática: El adjetivo,** *Assessment Program,* p. 51.

Cultura y lengua

Teach

- (15 min.) Have students read aloud **Cultura y lengua: Los mexicoamericanos,** pp. 94–95.

Wrap-Up

- (10 min.) Have students do Literature Link, *TRB,* p. 48.

OPTIONAL RESOURCES

- (20 min.) Have students do the first History Link, *TRB,* p. 48.
- (5 min.) Have groups of students do Thinking Critically, *TRB,* p. 48. ▲

Practice Options/Homework Suggestions

- Internet (go.hrw.com, Keyword: WN3 AMISTAD-CYL)
- Have students do Community Link, *TRB,* p. 49.
- Have students read **Cultura y lengua,** p. 96.

▲ = Advanced Learners ◆ = Slower Pace Learners ● = Special Learning Needs ■ = Heritage Speakers

Teacher's Name ______________________ Class ______________________ Date ____________

COLECCIÓN 2

Lazos de amistad

DAY 7 50-MINUTE LESSON PLAN

STANDARDS FOR FOREIGN LANGUAGE LEARNING: DAY 7

Cultura y lengua

Communication 1.1: Students engage in conversations, provide and obtain information, express feelings and emotions, and exchange opinions.

Connections 3.2: Students acquire information and recognize the distinctive viewpoints that are only available through the foreign language and its cultures.

Comparisons 4.1: Students demonstrate understanding of the nature of language through comparisons of the language studied and their own.

Comparisons 4.2: Students demonstrate understanding of the concept of culture through comparisons of the cultures studied and their own.

CORE INSTRUCTION

Warm-Up

- (5 min.) Have students discuss the Pair Work activity, *TRB,* pp. 48–49.

Cultura y lengua

Teach

- (15 min.) Have pairs of students do **Actividad,** p. 97.
- (15 min.) See **Cultura y lengua,** Teaching Suggestions, **Antes de ver** and **Mientras lo ves,** *Video Guide,* p. 12.
- (5 min.) Show **Cultura y lengua: Los mexicoamericanos,** *Video Program* (Videocassette 1).

Wrap-Up

- (10 min.) Play some Latin Grammy-winning music while discussing **Modismos y regionalismos,** p. 97. See the second Language Note, *TRB,* p. 49.

OPTIONAL RESOURCES

- (20 min.) Have students do the second History Link, *TRB,* p. 48. ▲
- (20 min.) Have students do the Community Link, *TRB,* p. 48.
- (10 min.) Have students do **Hoja de actividades 2,** *Video Guide,* p. 16.

Practice Options/Homework Suggestions

- Internet (go.hrw.com, Keyword: WN3 AMISTAD-CYL)
- Have students do Thinking Critically, *TRB,* p. 49. ▲
- *Advanced Placement Literature Preparation Book,* pp. 16–18 ▲

▲ = Advanced Learners ◆ = Slower Pace Learners ● = Special Learning Needs ■ = Heritage Speakers

Lazos de amistad

DAY 8 50-MINUTE LESSON PLAN

STANDARDS FOR FOREIGN LANGUAGE LEARNING: DAY 8

Cultura y lengua

Communication 1.1: Students engage in conversations, provide and obtain information, express feelings and emotions, and exchange opinions.

Cultures 2.2: Students demonstrate an understanding of the relationship between the products and perspectives of the culture studied.

Comparisons 4.1: Students demonstrate understanding of the nature of language through comparisons of the language studied and their own.

CORE INSTRUCTION

Warm-Up

- (5 min.) Have students think of other examples of "Spanglish." See **Modismos y regionalismos,** p. 97.

Cultura y lengua

Teach

- (25 min.) Have students do Re-entry, *TRB,* p. 49.
- (10 min.) Present **Así se dice,** p. 97.

Wrap-Up

- (10 min.) Have pairs of students use the expressions in **Así se dice,** p. 97, to discuss what they have learned about **Los mexicoamericanos.**

OPTIONAL RESOURCES

- (25 min.) Have students do research for Community Link, *TRB,* p. 49.
- (25 min.) Have students do **Así se dice,** *TRB,* p. 49. ▲

Practice Options/Homework Suggestions

- Internet (go.hrw.com, Keyword: WN3 AMISTAD-CYL)
- Have students read **Estrategias para leer,** pp. 98–99. ◆ ●
- Have pairs of students do research for **Cultura y lengua,** Teaching Suggestions, **Después de ver,** item 3, *Video Guide,* p. 12.
- Have students study for **Prueba de cultura.**

▲ = Advanced Learners ◆ = Slower Pace Learners ● = Special Learning Needs ■ = Heritage Speakers

Teacher's Name ______________________ Class ______________________ Date ____________

COLECCIÓN 2

Lazos de amistad

DAY 9 50-MINUTE LESSON PLAN

STANDARDS FOR FOREIGN LANGUAGE LEARNING: DAY 9

Cultura y lengua/Lectura

Communication 1.3: Students present information, concepts, and ideas to an audience of listeners or readers on a variety of topics.

Connections 3.1: Students reinforce and further their knowledge of other disciplines through the foreign language.

Communities 5.1: Students use the language both within and beyond the school setting.

CORE INSTRUCTION

Warm-Up

- (5 min.) Have students review **Cultura y lengua,** pp. 94–97.

Cultura y lengua

Assess

- (20 min.) Give **Prueba de cultura: Los mexicoamericanos,** *Assessment Program,* p. 55.

Lectura

Teach

- (10 min.) Have students do **Punto de partida,** p. 100. See **Punto de partida,** *TRB,* p. 51.
- (5 min.) Present **Elementos de literatura,** p. 100. See **Glosario de términos literarios,** p. R9.

Wrap-Up

- (10 min.) Have students give examples of irony from everyday life. See **Elementos de literatura,** *TRB,* p. 51.

OPTIONAL RESOURCES

- (10 min.) Have students do **Comparte tus ideas,** p. 100. See **Comparte tus ideas,** *TRB,* p. 51.
- (20 min.) Have students do **Para hispanohablantes,** *TRB,* p. 53. ■
- (20 min.) Have students do **Para angloparlantes,** *TRB,* p. 53.

Practice Options/Homework Suggestions

- Internet (go.hrw.com, Keyword: WN3 AMISTAD-LEC)
- Have students begin reading **"Una carta a Dios,"** pp. 102–103.
- Have students study **Vocabulario esencial, "Una carta a Dios,"** p. 139. ◆ ●

▲ = Advanced Learners ◆ = Slower Pace Learners ● = Special Learning Needs ■ = Heritage Speakers

Lazos de amistad

DAY 10 50-MINUTE LESSON PLAN

STANDARDS FOR FOREIGN LANGUAGE LEARNING: DAY 10

Lectura

Communication 1.1: Students engage in conversations, provide and obtain information, express feelings and emotions, and exchange opinions.

Communication 1.2: Students understand and interpret written and spoken language on a variety of topics.

Connections 3.1: Students reinforce and further their knowledge of other disciplines through the foreign language.

CORE INSTRUCTION

Warm-Up

- (5 min.) Have students read **Estrategias para leer,** p. 100.

Lectura

Teach

- (15 min.) Have students read aloud **"Una carta a Dios,"** pp. 104–105. See Techniques for Handling the Reading, *TRB*, p. 52.
- (5 min.) Have students read aloud **Conoce al escritor,** p. 105.
- (10 min.) Have students do **Repaso del texto,** p. 106. See **Repaso del texto,** *TRB*, p. 52.
- (5 min.) Have students do **Primeras impresiones,** p. 106. See **Primeras impresiones,** *TRB*, p. 52.

Wrap-Up

- (10 min.) Have students do **Conexiones con el texto,** p. 106.

OPTIONAL RESOURCES

- (10 min.) Read aloud **"Una carta a Dios,"** Summary, *TRB*, p. 51. ◆ ●
- (30 min.) Have students practice the Reading Strategy, *TRB*, pp. 249–250. ▲ ◆ ●

Practice Options/Homework Suggestions

- Internet (go.hrw.com, Keyword: WN3 AMISTAD-LEC)
- *Cuaderno de práctica,* Activities 1–3, pp. 27–28

▲ = Advanced Learners ◆ = Slower Pace Learners ● = Special Learning Needs ■ = Heritage Speakers

Lazos de amistad

DAY 11 50-MINUTE LESSON PLAN

STANDARDS FOR FOREIGN LANGUAGE LEARNING: DAY 11

Lectura

Communication 1.3: Students present information, concepts, and ideas to an audience of listeners or readers on a variety of topics.

Connections 3.1: Students reinforce and further their knowledge of other disciplines through the foreign language.

CORE INSTRUCTION

Warm-Up

- (5 min.) Have students do **Preguntas al texto,** p. 106.

Lectura

Teach

- (15 min.) Have students do **Cuaderno del escritor,** p. 107. See **Cuaderno del escritor,** *TRB,* p. 52.
- (20 min.) Have students write a letter of apology using the expressions in **Así se dice,** p. 106. See **Así se dice,** *TRB,* p. 52.

Wrap-Up

- (10 min.) Have volunteers share their letters of apology with the class.

OPTIONAL RESOURCES

- Have students study **Vocabulario adicional, "Una carta a Dios,"** *TRB,* p. 298. ◆ ●

Practice Options/Homework Suggestions

- Internet (go.hrw.com, Keyword: WN3 AMISTAD-LEC)
- *Cuaderno de práctica,* **Vocabulario adicional,** p. 139
- *Cuaderno de práctica,* Activity 4, p. 33

▲ = Advanced Learners ◆ = Slower Pace Learners ● = Special Learning Needs ■ = Heritage Speakers

Teacher's Name ______________________ Class ______________________ Date ______________

Lazos de amistad

DAY 12 50-MINUTE LESSON PLAN

STANDARDS FOR FOREIGN LANGUAGE LEARNING: DAY 12

Lectura

Communication 1.1: Students engage in conversations, provide and obtain information, express feelings and emotions, and exchange opinions.

Communication 1.2: Students understand and interpret written and spoken language on a variety of topics.

Communication 1.3: Students present information, concepts, and ideas to an audience of listeners or readers on a variety of topics.

Communities 5.1: Students use the language both within and beyond the school setting.

CORE INSTRUCTION

Warm-Up

- (5 min.) Have students begin **Redacción creativa/Teatro,** p. 107, by picking one of the two situations listed.

Lectura

Teach

- (15 min.) Have students do **Redacción creativa/Teatro,** p. 107. See **Redacción creativa/Teatro,** *TRB,* p. 53.
- (10 min.) Have students do **Vocabulario en contexto,** Activities C and D, p. 116. See **Vocabulario en contexto,** *TRB,* p. 60. You may wish to use Audio CD 1, Track 12.
- (10 min.) Have students do Additional Practice, *TRB,* p. 60.

Wrap-Up

- (10 min.) Have volunteers read their telephone conversations to the class.

OPTIONAL RESOURCES

- (20 min.) Have students do **Redacción creativa,** p. 107. See **Redacción creativa,** *TRB,* pp. 52–53. ◆ ●

Practice Options/Homework Suggestions

- Internet (go.hrw.com, Keyword: WN3 AMISTAD-LEC)
- Have students study for **Prueba de lectura.**
- *Advanced Placement Literature Preparation Book,* pp. 19–39 ▲

▲ = Advanced Learners ◆ = Slower Pace Learners ● = Special Learning Needs ■ = Heritage Speakers

Teacher's Name ______________________ Class ______________________ Date ____________

Lazos de amistad

DAY 13 50-MINUTE LESSON PLAN

STANDARDS FOR FOREIGN LANGUAGE LEARNING: DAY 13

Lectura

Communication 1.2: Students understand and interpret written and spoken language on a variety of topics.

Connections 3.1: Students reinforce and further their knowledge of other disciplines through the foreign language.

Comparisons 4.1: Students demonstrate understanding of the nature of language through comparisons of the language studied and their own.

CORE INSTRUCTION

Warm-Up

- (5 min.) Have students review **"Una carta a Dios."**

Lectura
Assess

- (30 min.) Give **Prueba de lectura: "Una carta a Dios,"** *Assessment Program,* pp. 45–46.

Lectura
Teach

- (10 min.) Present the introduction of **Elementos de literatura,** p. 110. Do Getting Started, *TRB,* p. 55.

Wrap-Up

- (5 min.) Present the information in Language Note, *TRB,* p. 55.

OPTIONAL RESOURCES

- (15 min.) Have students read **A leer por tu cuenta: "La muralla,"** pp. 112–113. See **"La muralla,"** *TRB,* pp. 57–58. ▲
- (5 min.) Show the video **Lectura: "La muralla"** (Videocassette 1) or have students listen to Audio CD 3, Track 4. ▲
- (30 min.) See **Lectura,** Teaching Suggestions, *Video Guide,* p. 11. ▲
- (15 min.) Have students do **Hoja de actividades 1,** *Video Guide,* p. 15. ▲
- (15 min.) Have students practice the Reading Strategy, *TRB,* pp. 251–252. ◆ ●

Practice Options/Homework Suggestions

- Internet (go.hrw.com, Keyword: WN3 AMISTAD-LEC)
- Have students read the rest of **Elementos de literatura,** pp. 110–111. ▲
- *Cuaderno de práctica,* pp. 29–31

▲ = Advanced Learners ◆ = Slower Pace Learners ● = Special Learning Needs ■ = Heritage Speakers

Teacher's Name ______________________ Class ____________________ Date __________

COLECCIÓN 2

Lazos de amistad

DAY 14 50-MINUTE LESSON PLAN

STANDARDS FOR FOREIGN LANGUAGE LEARNING: DAY 14

Lectura/Vocabulario

Communication 1.3: Students present information, concepts, and ideas to an audience of listeners or readers on a variety of topics.

Connections 3.1: Students reinforce and further their knowledge of other disciplines through the foreign language.

Comparisons 4.1: Students demonstrate understanding of the nature of language through comparisons of the language studied and their own.

CORE INSTRUCTION

Warm-Up

- (5 min.) Have students review **Elementos de literatura,** p. 111.

Lectura/Vocabulario

Teach

- (20 min.) Have small groups do the Group Work activity (Applying the Element), *TRB,* p. 55.
- (10 min.) Present **Mejora tu vocabulario: Anglicismos,** p. 116. See **Mejora tu vocabulario: Anglicismos,** *TRB,* p. 61.
- (10 min.) Have students do Activity F, p. 118.

Wrap-Up

- (5 min.) Have volunteers read their definitions to the class.

OPTIONAL RESOURCES

- (30 min.) Have students discuss Applying the Element, *TRB,* p. 55.
- (20 min.) Have students do Community Link, *TRB,* pp. 55–56.
- (20 min.) Have students do Thinking Critically, *TRB,* p. 56. ▲
- (15 min.) Have students do Art Link, *TRB,* p. 56.
- (15 min.) Have students do Additional Practice, *TRB,* p. 56. ◆ ●

Practice Options/Homework Suggestions

- Internet (go.hrw.com, Keyword: WN3 AMISTAD-LEC)
- Have students do Activity E, p. 117.
- *Video Guide,* **Hoja de actividades 1, "La muralla,"** p. 15

Assessment Options

- *Assessment Program,* **Prueba de lectura: "La muralla,"** pp. 47–48

▲ = Advanced Learners ◆ = Slower Pace Learners ● = Special Learning Needs ■ = Heritage Speakers

Lazos de amistad

DAY 15 50-MINUTE LESSON PLAN

STANDARDS FOR FOREIGN LANGUAGE LEARNING: DAY 15

Vocabulario

Communication 1.2: Students understand and interpret written and spoken language on a variety of topics.

Comparisons 4.1: Students demonstrate understanding of the nature of language through comparisons of the language studied and their own.

CORE INSTRUCTION

Warm-Up

- (5 min.) Have students review **Anglicismos,** pp. 116–117.

Vocabulario

Teach

- (15 min.) Have students do Activity H, p. 118. See Activity H, *TRB,* p. 62.
- (20 min.) Have students do Activity I, p. 62. See Activity I, *TRB,* p. 62.

Wrap-Up

- (10 min.) Review answers for Activity I, p. 62.

OPTIONAL RESOURCES

- (20 min.) Have students do **Para hispanohablantes,** *TRB,* p. 62. ■
- (10 min.) Have students do **Para angloparlantes,** *TRB,* p. 62.
- (15 min.) Have students do Activity G, p. 118. See Activity G, *TRB,* p. 61.
- (20 min.) Have students do Cooperative Learning, *TRB,* p. 62. ▲ ●

Practice Options/Homework Suggestions

- *Cuaderno de práctica,* Activities 6–8, pp. 34–35
- Have students study for **Vocabulario esencial, Mejora tu vocabulario,** p. 139.
- Have students study for **Prueba de vocabulario.**

Assessment Options

- *Assessment Program,* **Prueba de vocabulario adicional,** p. 257

▲ = Advanced Learners ◆ = Slower Pace Learners ● = Special Learning Needs ■ = Heritage Speakers

Teacher's Name ______________________ Class ______________________ Date ____________

COLECCIÓN 2

Lazos de amistad

DAY 16 50-MINUTE LESSON PLAN

STANDARDS FOR FOREIGN LANGUAGE LEARNING: DAY 16

Vocabulario/Gramática

Communication 1.2: Students understand and interpret written and spoken language on a variety of topics.

Communication 1.3: Students present information, concepts, and ideas to an audience of listeners or readers on a variety of topics.

Comparisons 4.1: Students demonstrate understanding of the nature of language through comparisons of the language studied and their own.

CORE INSTRUCTION

Warm-Up

- (5 min.) Have students review **Vocabulario,** pp. 115–119.

Vocabulario

Assess

- (20 min.) Give **Prueba de vocabulario,** *Assessment Program,* pp. 49–50.

Gramática

Teach

- (10 min.) Present **El adverbio,** p. 122. See **El adverbio,** *TRB,* p. 64.
- (10 min.) Have students do Activities D and E, pp. 122–123. See Activities D and E, *TRB,* p. 64.

Wrap-Up

- (5 min.) Have pairs of student do the first Pair Work activity, *TRB,* p. 64.

OPTIONAL RESOURCES

- (10 min.) Have students do Activity G, p. 124.
- (15 min.) Have students do the second Pair Work activity, *TRB,* p. 64.

Practice Options/Homework Suggestions

- *Cuaderno de práctica,* Activities 4–6, pp. 37–38

▲ = Advanced Learners ◆ = Slower Pace Learners ● = Special Learning Needs ■ = Heritage Speakers

Teacher's Name ______________________ Class ____________________ Date __________

COLECCIÓN **2**

Lazos de amistad

DAY 17 50-MINUTE LESSON PLAN

STANDARDS FOR FOREIGN LANGUAGE LEARNING: DAY 17

Gramática

Communication 1.1: Students engage in conversations, provide and obtain information, express feelings and emotions, and exchange opinions.

Communication 1.2: Students understand and interpret written and spoken language on a variety of topics.

Comparisons 4.1: Students demonstrate understanding of the nature of language through comparisons of the language studied and their own.

CORE INSTRUCTION

Warm-Up

- (5 min.) Have students review **El adverbio,** pp. 122–124.

Gramática

Teach

- (5 min.) Have students do Activity F, p. 123.
- (15 min.) Present **El comparativo,** p. 124. See **El comparativo** and Language Note, *TRB,* pp. 64–65.
- (15 min.) Have students do Activities H and I, pp. 124–125.

Wrap-Up

- (10 min.) Have students form groups to do Extension, *TRB,* p. 65.

OPTIONAL RESOURCES

- (15 min.) Have students do Activity K, p. 126.
- (20 min.) Have students do Group Work, *TRB,* p. 65. ▲
- (20 min.) Have students do **Para hispanohablantes,** *TRB,* p. 66. ■

Practice Options/Homework Suggestions

- *Cuaderno de práctica,* Activities 7–10, pp. 39–40
- Have students study for **Prueba de gramática.**
- *Advanced Placement Literature Preparation Book,* pp. 40–42 ▲

▲ = Advanced Learners ◆ = Slower Pace Learners ● = Special Learning Needs ■ = Heritage Speakers

Teacher's Name ______________________ Class ______________________ Date ____________

COLECCIÓN 2

Lazos de amistad

DAY 18 50-MINUTE LESSON PLAN

STANDARDS FOR FOREIGN LANGUAGE LEARNING: DAY 18

Gramática

Communication 1.2: Students understand and interpret written and spoken language on a variety of topics.

Comparisons 4.1: Students demonstrate understanding of the nature of language through comparisons of the language studied and their own.

CORE INSTRUCTION

Warm-Up

- (5 min.) Have students review **El adverbio,** pp. 122–124, and **El comparativo,** pp. 124–126.

Gramática

Assess

- (30 min.) Give **Prueba de gramática: El adverbio** and **Prueba de gramática: El comparativo,** *Assessment Program,* pp. 52 and 53.
- (5 min.) Present **Comparación y contraste,** p. 126. See **Comparación y contraste,** *TRB,* pp. 65–66.
- (5 min.) Have students do Activity A, p. 127.

Wrap-Up

- (5 min.) Have students do Activity B, p. 127.

OPTIONAL RESOURCES

- (10 min.) Have students do **Para angloparlantes,** *TRB,* p. 66.

Practice Options/Homework Suggestions

- Have students read **Panorama cultural,** pp. 108–109.
- *Cuaderno de práctica,* Activities 11–13, p. 41
- Have students study for **Prueba de comparación y contraste.**

▲ = Advanced Learners ◆ = Slower Pace Learners ● = Special Learning Needs ■ = Heritage Speakers

Lazos de amistad

DAY 19 50-MINUTE LESSON PLAN

STANDARDS FOR FOREIGN LANGUAGE LEARNING: DAY 19

Gramática/Panorama cultural

Communication 1.2: Students understand and interpret written and spoken language on a variety of topics.

Cultures 2.1: Students demonstrate an understanding of the relationship between the practices and perspectives of the culture studied.

Connections 3.2: Students acquire information and recognize the distinctive viewpoints that are only available through the foreign language and its cultures.

Comparisons 4.2: Students demonstrate understanding of the concept of culture through comparisons of the cultures studied and their own.

CORE INSTRUCTION

Warm-Up

- (5 min.) Have students review **Comparación y contraste,** pp. 126–127.

Gramática

Assess

- (20 min.) Give **Prueba de comparación y contraste,** *Assessment Program,* p. 54.

Panorama cultural

Teach

- (10 min.) Have students discuss the introduction of **Panorama cultural,** p. 108. See **Panorama cultural,** Summary and Presentation, *TRB,* p. 54.
- (5 min.) Play Audio CD 1, Tracks 9–10, and have students listen to the interviews.

Wrap-Up

- (10 min.) Have students do Activities A and B, p. 109.

OPTIONAL RESOURCES

- (20 min.) Have students do Culture Link, *TRB,* p. 54. ▲ ■
- (20 min.) Have students do **Panorama cultural,** Teaching Suggestions, **Antes de ver,** *Video Guide,* p. 13.
- (10 min.) Have students do Activity D, p. 109. You may use Audio CD 1, Track 11. See script, *TRB,* p. 54.
- (20 min.) Have students do **Panorama cultural,** Teaching Suggestions, **Después de ver,** *Video Guide,* p. 13.
- (15 min.) Show **Panorama cultural,** *Video Program* (Videocassette 1). See **Panorama cultural,** Teaching Suggestions, **Mientras lo ves,** *Video Guide,* p. 13.
- (15 min.) Have students do **Hoja de actividades 3,** *Video Guide,* p. 17.

Practice Options/Homework Suggestions

- Have students read **Comunidad y oficio,** p. 114.
- Have students do Community Link, *TRB,* p. 54. ▲ ■

▲ = Advanced Learners ◆ = Slower Pace Learners ● = Special Learning Needs ■ = Heritage Speakers

Teacher's Name ______________________ Class ______________________ Date ____________

COLECCIÓN 2

Lazos de amistad

DAY 20 50-MINUTE LESSON PLAN

STANDARDS FOR FOREIGN LANGUAGE LEARNING: DAY 20

Comunidad y oficio

Communication 1.1: Students engage in conversations, provide and obtain information, express feelings and emotions, and exchange opinions.

Communication 1.2: Students understand and interpret written and spoken language on a variety of topics.

Communication 1.3: Students present information, concepts, and ideas to an audience of listeners or readers on a variety of topics.

Communities 5.1: Students use the language both within and beyond the school setting.

Communities 5.2: Students show evidence of becoming life-long learners by using the language for personal enjoyment and enrichment.

CORE INSTRUCTION

Warm-Up

- (5 min.) Have students do Activity C, p. 109.

Comunidad y oficio

Teach

- (5 min.) Present **Comunidad y oficio,** p. 114. See Getting Started, *TRB,* p. 59.
- (10 min.) Have students read and discuss **Comunidad y oficio,** p. 114.
- (10 min.) Have students do **Comunidad y oficio,** Teaching Suggestions, **Antes de ver,** *Video Guide,* p. 14.
- (10 min.) Show **Comunidad y oficio,** *Video Program* (Videocassette 1). See **Comunidad y oficio,** Teaching Suggestions, **Mientras lo ves,** *Video Guide,* p. 14.

Wrap-Up

- (10 min.) Have students do **Comunidad y oficio,** Teaching Suggestions, **Después de ver,** item two, *Video Guide,* p. 18.

OPTIONAL RESOURCES

- (30 min.) Have students do **Investigaciones,** Activity B, p. 114. See **Investigaciones,** *TRB,* p. 59. ▲
- (30 min.) Have students do Community Link, *TRB,* p. 59.
- (30 min.) Have students do the first Thinking Critically activity, *TRB,* p. 59. ▲
- (15 min.) Have students do the second Thinking Critically activity, *TRB,* p. 59. ▲
- (5 min.) Have students do the Culture Note, *TRB,* p. 59.
- (5 min.) Have students do the Language Note, *TRB,* p. 59.

Practice Options/Homework Suggestions

- Internet (go.hrw.com, Keyword: WN3 AMISTAD-CYO)
- Have students do **Investigaciones,** Activity A, p. 114. ▲

▲ = Advanced Learners ◆ = Slower Pace Learners ● = Special Learning Needs ■ = Heritage Speakers

Teacher's Name ______________________ Class ______________________ Date ____________

COLECCIÓN 2

Lazos de amistad

DAY 21 50-MINUTE LESSON PLAN

STANDARDS FOR FOREIGN LANGUAGE LEARNING: DAY 21

Ortografía

Communication 1.2: Students understand and interpret written and spoken language on a variety of topics.

Comparisons 4.1: Students demonstrate understanding of the nature of language through comparisons of the language studied and their own.

CORE INSTRUCTION

Warm-Up

- (5 min.) Have students read the introduction of **Los sonidos** ***/r/*** **y** ***/rr/,*** p. 128.

Ortografía

Teach

- (10 min.) Present **Los sonidos** ***/r/*** **y** ***/rr/,*** p. 128. See **Los sonidos** ***/r/*** **y /rr/** and Language Note, *TRB,* p. 67.
- (10 min.) Have students do Activities A and B, pp. 128–129.
- (10 min.) Present **El sonido** ***/y/,*** p. 129. See **El sonido** ***/y/*** and Language Note, *TRB,* p. 68.
- (10 min.) Have students do Activities C and D, p. 130.

Wrap-Up

- (5 min.) Review **r, rr, y,** and **ll** by saying words aloud and having students say what letter it is. For example, **raya (y), cierro (rr),** and so on.

OPTIONAL RESOURCES

- (5 min.) Go over **¡Ojo!** with students, pp. 128 and 129.
- (10 min.) Have students do the Pair Work activity, *TRB,* p. 67. ▲
- (10 min.) Have students do Extension, *TRB,* pp. 67–68. ▲
- (15 min.) Have students do second Extension, *TRB,* pp. 68. ▲
- (20 min.) Have students do **Ampliación, Hoja de práctica 2-A, Confusión entre las terminaciones -ío/-ía e -illo/-illa,** *TRB,* p. 282. ◆ ●

Practice Options/Homework Suggestions

- *Cuaderno de práctica,* Activities 1–5, pp. 42–43
- *Cuaderno de práctica,* **Ampliación, Hoja de práctica 2-A,** p. 140

▲ = Advanced Learners ◆ = Slower Pace Learners ● = Special Learning Needs ■ = Heritage Speakers

Lazos de amistad

DAY 22 50-MINUTE LESSON PLAN

STANDARDS FOR FOREIGN LANGUAGE LEARNING: DAY 22

Ortografía

Communication 1.2: Students understand and interpret written and spoken language on a variety of topics.

Communication 1.3: Students present information, concepts, and ideas to an audience of listeners or readers on a variety of topics.

Comparisons 4.1: Students demonstrate understanding of the nature of language through comparisons of the language studied and their own.

CORE INSTRUCTION

Warm-Up

- (5 min.) Have students review **Los sonidos** ***/r/*** **y** ***/rr/*** and **El sonido** ***/y/,*** pp. 128–129.

Ortografía

Teach

- (10 min.) Present **La acentuación,** pp. 130–131. See **La acentuación,** *TRB,* p. 68.
- (10 min.) Have students do Activities E and F, p. 131.
- (20 min.) Give **Dictado,** Activities A and B, p. 131. You may wish to use Audio CD 1, Tracks 13–14. See scripts, *TRB,* p. 69.

Wrap-Up

- (5 min.) Review rules for accent marks.

OPTIONAL RESOURCES

- (15 min.) Have students do Extension, *TRB,* p. 68, or Extension, *TRB,* p. 69. ▲
- (20 min.) Have students do **Para hispanohablantes,** *TRB,* p. 69. ■
- (20 min.) Have students do **Para angloparlantes,** *TRB,* p. 69.

Practice Options/Homework Suggestions

- *Cuaderno de práctica,* Activities 6–8, pp. 43–44
- Have students study for **Prueba de ortografía.**
- *Advanced Placement Literature Preparation Book,* pp. 43–46 ▲

Assessment Options

- *Assessment Program,* **Prueba de ortografía 2-A: Confusión entre las terminaciones -ío/ía e -illo/illa,** p. 258

▲ = Advanced Learners ◆ = Slower Pace Learners ● = Special Learning Needs ■ = Heritage Speakers

Lazos de amistad

DAY 23 50-MINUTE LESSON PLAN

STANDARDS FOR FOREIGN LANGUAGE LEARNING: DAY 23

Ortografía/Taller del escritor

Communication 1.1: Students engage in conversations, provide and obtain information, express feelings and emotions, and exchange opinions.

Communication 1.2: Students understand and interpret written and spoken language on a variety of topics.

Connections 3.1: Students reinforce and further their knowledge of other disciplines through the foreign language.

CORE INSTRUCTION

Warm-Up

- (5 min.) Have students review **Letra y sonido** and **La acentuación,** pp. 128–131.

Ortografía

Assess

- (20 min.) Give **Prueba de ortografía,** *Assessment Program,* p. 56.

Taller del escritor

Teach

- (5 min.) Have students review their work on the portfolio suggestions after each reading selection. See **Opciones: Prepara tu portafolio,** pp. 93 and 107.
- (10 min.) Present **Semblanza,** p. 132. See **Semblanza,** Presenting the Workshop, *TRB,* p. 70.
- (5 min.) Introduce **Antes de escribir: Lluvia de ideas,** p. 132. See **Antes de escribir,** *TRB,* p. 70.

Wrap-Up

- (5 min.) Model brainstorming for a **semblanza** on the chalkboard.

OPTIONAL RESOURCES

- (20 min.) Have students read and discuss **Escritura libre,** p. 132, and **Idea principal,** p. 133.

Practice Options/Homework Suggestions

- Have students do **Escritura libre,** p. 132.

▲ = Advanced Learners ◆ = Slower Pace Learners ● = Special Learning Needs ■ = Heritage Speakers

Teacher's Name ______________________ Class ______________________ Date ____________

COLECCIÓN 2

Lazos de amistad

DAY 24 50-MINUTE LESSON PLAN

STANDARDS FOR FOREIGN LANGUAGE LEARNING: DAY 24

Taller del escritor

Communication 1.1: Students engage in conversations, provide and obtain information, express feelings and emotions, and exchange opinions.

Communication 1.2: Students understand and interpret written and spoken language on a variety of topics.

Communication 1.3: Students present information, concepts, and ideas to an audience of listeners or readers on a variety of topics.

CORE INSTRUCTION

Warm-Up

- (5 min.) Have students study the chart, shown in **Investigación,** pp. 132–133.

Taller del escritor

Teach

- (15 min.) Have students, either individually or in groups, do a chart for their own narrative, pp. 132–133.
- (10 min.) Present **Objetivo y público,** p. 133.
- (15 min.) Have students do **Reúne detalles,** pp. 133–134.

Wrap-Up

- (5 min.) Review **Detalles para una semblanza,** p. 134.

OPTIONAL RESOURCES

- (20 min.) Have students compare the authors in this collection and their works, pp. 90 and 105. ▲

Practice Options/Homework Suggestions

- Have students read **El borrador,** p. 134.

▲ = Advanced Learners ◆ = Slower Pace Learners ● = Special Learning Needs ■ = Heritage Speakers

Teacher's Name ______________________ Class ______________________ Date ____________

Lazos de amistad

DAY 25 50-MINUTE LESSON PLAN

STANDARDS FOR FOREIGN LANGUAGE LEARNING: DAY 25

Taller del escritor

Communication 1.1: Students engage in conversations, provide and obtain information, express feelings and emotions, and exchange opinions.

Communication 1.2: Students understand and interpret written and spoken language on a variety of topics.

Communities 5.1: Students use the language both within and beyond the school setting.

CORE INSTRUCTION

Warm-Up

- (5 min.) Have students review **Esquema para una semblanza,** p. 134.

Taller del escritor

Teach

- (5 min.) Present **El borrador,** p. 134. See **El borrador,** *TRB,* p. 70.
- (5 min.) Present **Esquema para una semblanza,** p. 134.
- (30 min.) Have students write the first draft of **Una semblanza.**

Wrap-Up

- (5 min.) Have students find a Spanish-language magazine or newspaper article that describes a person, and have them discuss it.

OPTIONAL RESOURCES

- (15 min.) Have students work on their outlines in class. See **Esquema para una semblanza,** p. 134.

Practice Options/Homework Suggestion

- Have students revise their drafts.

▲ = Advanced Learners ◆ = Slower Pace Learners ● = Special Learning Needs ■ = Heritage Speakers

Teacher's Name ______________________ Class ______________________ Date ____________

COLECCIÓN 2

Lazos de amistad

DAY 26 50-MINUTE LESSON PLAN

STANDARDS FOR FOREIGN LANGUAGE LEARNING: DAY 26

Taller del escritor

Communication 1.1: Students engage in conversations, provide and obtain information, express feelings and emotions, and exchange opinions.

Communication 1.3: Students present information, concepts, and ideas to an audience of listeners or readers on a variety of topics.

Communities 5.1: Students use the language both within and beyond the school setting.

CORE INSTRUCTION

Warm-Up

- (5 min.) Have students read the expressions in **Así se dice,** p. 135.

Taller del escritor

Teach

- (20 min.) Have students work in pairs to evaluate each other's rough draft. Have them use the questions in **Evaluación y revisión,** Activity 1, pp. 134–135, and the expressions in **Así se dice,** p. 135. See **Evaluación y revisión,** *TRB,* p. 70.
- (20 min.) Have students evaluate their work using **Pautas de evaluación** and **Técnicas de revisión,** p. 134. See **Evaluación y revisión,** *TRB,* p. 70.

Wrap-Up

- (5 min.) Have students begin revising their drafts.

OPTIONAL RESOURCES

- (20 min.) Have students do the Pair Work activity, *TRB,* p. 71. ◆ ●
- (10 min.) Have students do Additional Practice, *TRB,* p. 70. ◆ ●

Practice Options/Homework Suggestions

- Have students use the suggestions in **Autoevaluación,** p. 135, to revise their drafts.

▲ = Advanced Learners ◆ = Slower Pace Learners ● = Special Learning Needs ■ = Heritage Speakers

Teacher's Name ______________ Class ______________ Date ______________

COLECCIÓN 2

Lazos de amistad

DAY 27 50-MINUTE LESSON PLAN

STANDARDS FOR FOREIGN LANGUAGE LEARNING: DAY 27

Taller del escritor

Communication 1.1: Students engage in conversations, provide and obtain information, express feelings and emotions, and exchange opinions.

Communication 1.3: Students present information, concepts, and ideas to an audience of listeners or readers on a variety of topics.

Connections 3.1: Students reinforce and further their knowledge of other disciplines through the foreign language.

Communities 5.2: Students show evidence of becoming life-long learners by using the language for personal enjoyment and enrichment.

CORE INSTRUCTION

Warm-Up

- (5 min.) Have students begin reading the **Modelos** on pp. 135–136.

Taller del escritor

Teach

- (20 min.) Have students do **Corrección de pruebas,** p. 136. See **Corrección de pruebas,** *TRB,* p. 71.
- (15 min.) Have students do one of the activities in **Publicación,** p. 136.
- (5 min.) Have students do **Reflexión,** p. 136. See **Reflexión,** *TRB,* p. 71.

Wrap-Up

- (5 min.) Do Closure, *TRB,* p. 71.

OPTIONAL RESOURCES

- (10 min.) Have students reflect on their work using the expressions in **Así se dice,** p. 136.
- (20 min.) Have students do **Publicación,** *TRB,* p. 71. ▲
- (30 min.) Have students do Enrichment, *TRB,* p. 71.

Practice Options/Homework Suggestions

- Have students do **A ver si puedo…,** pp. 137–138.

▲ = Advanced Learners ◆ = Slower Pace Learners ● = Special Learning Needs ■ = Heritage Speakers

Teacher's Name ______________________ Class ______________________ Date ____________

Lazos de amistad

DAY 28 50-MINUTE LESSON PLAN

STANDARDS FOR FOREIGN LANGUAGE LEARNING: DAY 28

A ver si puedo…

Communication 1.2: Students understand and interpret written and spoken language on a variety of topics.

Communication 1.3: Students present information, concepts, and ideas to an audience of listeners or readers on a variety of topics.

Connections 3.1: Students reinforce and further their knowledge of other disciplines through the foreign language.

Connections 3.2: Students acquire information and recognize the distinctive viewpoints that are only available through the foreign language and its cultures.

CORE INSTRUCTION

Warm-Up

- (5 min.) Have students review the objectives listed on the Collection Opener, p. 80.

A ver si puedo…

Review

- (10 min.) Have students do **Lectura,** Activities A and B, p. 137.
- (10 min.) Have students do **Cultura,** Activity C, p. 137.
- (10 min.) Have students do **Comunicación,** Activities D, E, F, and G, pp. 137–138.
- (10 min.) Have students do **Escritura,** Activities H, I, and J, p. 138.

Wrap-Up

- (5 min.) Answer any questions about either of the two chapter exams.

OPTIONAL RESOURCES

- (35 min.) Have students read **Enlaces Literarios: El soneto del Siglo de Oro,** pp. 140–144. See *TRB,* pp. 74–77. ▲
- (20 min.) Have students do **Comprensión del texto,** p. 145. See *TRB,* p. 76. ▲
- (25 min.) Have students do **Análisis del texto,** p. 145. See *TRB,* pp. 76–77. ▲
- (25 min.) Have students do **Más allá del texto,** p. 145. See *TRB,* p. 77. ▲
- (25 min.) Have students select from activities suggested in the *TRB* for **El soneto del Siglo de Oro:** Art Link, Literature Link, Music Link, pp. 75–76. ▲

Practice Options/Homework Suggestions

- Have students study for the **Examen de lengua.**

▲ = Advanced Learners ◆ = Slower Pace Learners ● = Special Learning Needs ■ = Heritage Speakers

Lazos de amistad

DAY 29 50-MINUTE LESSON PLAN

CORE INSTRUCTION

Assess

- (50 min.) Give **Colección 2 Examen de lengua,** *Assessment Program,* pp. 63–70.

OPTIONAL RESOURCES

- (50 min.) Give **Examen de lectura: "Cadena rota," "Naranjas," y "Una carta a Dios,"** *Assessment Program,* pp. 57–62. To allow students more time to take the exam, either **Examen** may be given over two class periods.

Practice Options/Homework Suggestions

- Have students study for the **Examen de lectura: "Cadena rota," "Naranjas" y "Una carta a Dios."**

Assessment Options

- *Assessment Program,* Performance Assessment, p. 296

▲ = Advanced Learners ◆ = Slower Pace Learners ● = Special Learning Needs ■ = Heritage Speakers

Teacher's Name ______________________ Class ______________________ Date ____________

COLECCIÓN

Lazos de amistad

DAY 30 50-MINUTE LESSON PLAN

CORE INSTRUCTION

Assess

- (50 min.) Give **Examen de lectura: "Cadena rota," "Naranjas" y "Una carta a Dios,"** *Assessment Program,* pp. 57–62.

OPTIONAL RESOURCES

- (50 min.) Give **Colección 2 Examen de lengua,** *Assessment Program,* pp. 63–70. To allow students more time to take the exam, either **Examen** may be given over two class periods.

Practice Options/Homework Suggestions

- Internet (go.hrw.com, Keyword: WN3 AMISTAD)

Assessment Options

- *Assessment Program,* Performance Assessment, p. 296

▲ = Advanced Learners ◆ = Slower Pace Learners ● = Special Learning Needs ■ = Heritage Speakers

Teacher's Name ______________________ Class ____________________ Date __________

COLECCIÓN 3

El frágil medio ambiente

DAY 1 50-MINUTE LESSON PLAN

STANDARDS FOR FOREIGN LANGUAGE LEARNING: DAY 1

Lectura

Communication 1.1: Students engage in conversations, provide and obtain information, express feelings and emotions, and exchange opinions.

Communication 1.2: Students understand and interpret written and spoken language on a variety of topics.

Communication 1.3: Students present information, concepts, and ideas to an audience of listeners or readers on a variety of topics.

Connections 3.1: Students reinforce and further their knowledge of other disciplines through the foreign language.

Communities 5.1: Students use the language both within and beyond the school setting.

CORE INSTRUCTION

Warm-Up

- (5 min.) Have students read the objectives on the Collection Opener, p. 146. See Collection Overview, *TRB,* p. 82.

Lectura

Teach

- (15 min.) Have students answer questions in **Punto de partida,** p. 148. See **Punto de partida,** *TRB,* p. 83.
- (25 min.) Have students begin reading **"La fiesta del árbol,"** pp. 149–151. See Techniques for Handling the Reading, *TRB,* p. 83.

Wrap-Up

- (5 min.) See Additional Practice, *TRB,* p. 83.

OPTIONAL RESOURCES

- (10 min.) See Presentation Suggestions, item one, *TRB,* p. 82.
- (10 min.) See Presentation Suggestions, item two, *TRB,* p. 82.
- (10 min.) See **Vocabulario en contexto,** Group Work, *TRB,* p. 101.
- (10 min.) See **Estrategias para leer,** *TRB,* p. 83. ◆ ●
- (10 min.) Read aloud **"La fiesta del árbol,"** Summary, *TRB,* p. 83. ◆ ●

Practice Options/Homework Suggestions

- Internet (go.hrw.com, Keyword: WN3 AMBIENTE-LEC)
- Have students finish reading **"La fiesta del árbol,"** pp. 149–151, including **Conoce a la escritora,** p. 151.
- Have students practice the Reading Strategy, *TRB,* pp. 253–254. ◆ ●
- Have students study **Vocabulario esencial, "La fiesta del árbol,"** p. 215. ◆ ●
- *Advanced Placement Literature Preparation Book,* pp. 47–50 ▲

▲ = Advanced Learners ◆ = Slower Pace Learners ● = Special Learning Needs ■ = Heritage Speakers

Teacher's Name ______________ Class ______________ Date ______________

COLECCIÓN 3

El frágil medio ambiente

DAY 2 50-MINUTE LESSON PLAN

STANDARDS FOR FOREIGN LANGUAGE LEARNING: DAY 2

Lectura/Vocabulario

Communication 1.1: Students engage in conversations, provide and obtain information, express feelings and emotions, and exchange opinions.

Communication 1.2: Students understand and interpret written and spoken language on a variety of topics.

Connections 3.1: Students reinforce and further their knowledge of other disciplines through the foreign language.

Communities 5.1: Students use the language both within and beyond the school setting.

CORE INSTRUCTION

Warm-Up

- (5 min.) Have students read the definition of **ensayo** in ***Glosario de términos literarios,*** p. R8.

Lectura/Vocabulario

Teach

- (10 min.) Have students do **Vocabulario en contexto,** Activities A and B, p. 189. See Activities A and B, *TRB,* p. 101.
- (25 min.) Show **"La fiesta del árbol,"** *Video Program* (Videocassette 1). See Teaching Suggestions, **Mientras lo ves,** *Video Guide,* p. 20.

Wrap-Up

- (10 min.) Have students do **Hoja de actividades 1, Después de ver,** *Video Guide,* p. 24.

OPTIONAL RESOURCES

- (30 min.) Have students do the Pair Work activity, *TRB,* p. 101.
- (20 min.) See Teaching Suggestions, **Después de ver,** *Video Guide,* p. 20.
- (10 min.) Have students do **Hoja de actividades 1, Mientras lo ves,** *Video Guide,* p. 24.
- (20 min.) Have students do **Para hispanohablantes,** *TRB,* p. 85. ■
- (20 min.) Have students do **Para angloparlantes,** *TRB,* p. 85.

Practice Options/Homework Suggestions

- Internet (go.hrw.com, Keyword: WN3 AMBIENTE-LEC)
- Have students study **Vocabulario adicional, "La fiesta del árbol,"** *TRB,* p. 299. ◆ ●
- *Cuaderno de práctica,* Activities 1–3, pp. 45–46

▲ = Advanced Learners ◆ = Slower Pace Learners ● = Special Learning Needs ■ = Heritage Speakers

Teacher's Name ______________________ Class ____________________ Date __________

COLECCIÓN 3

El frágil medio ambiente

DAY 3 50-MINUTE LESSON PLAN

STANDARDS FOR FOREIGN LANGUAGE LEARNING: DAY 3

Lectura

Communication 1.1: Students engage in conversations, provide and obtain information, express feelings and emotions, and exchange opinions.

Communication 1.2: Students understand and interpret written and spoken language on a variety of topics.

Communication 1.3: Students present information, concepts, and ideas to an audience of listeners or readers on a variety of topics.

Connections 3.1: Students reinforce and further their knowledge of other disciplines through the foreign language.

Communities 5.1: Students use the language both within and beyond the school setting.

Communities 5.2: Students show evidence of becoming life-long learners by using the language for personal enjoyment and enrichment.

CORE INSTRUCTION

Warm-Up

- (5 min.) Have students discuss **Primeras impresiones,** p. 152.

Lectura

Teach

- (10 min.) Have students do **Interpretaciones del texto,** p. 152. See **Interpretaciones del texto,** *TRB,* p. 84.
- (5 min.) Have students do **Preguntas al texto,** p. 152.
- (25 min.) In pairs, have students do **Repaso del texto,** p. 152. See **Repaso del texto,** *TRB,* pp. 83–84. Have students use expressions from **Así se dice,** p. 152.

Wrap-Up

- (5 min.) Have volunteers model their interviews from **Repaso del texto.**

OPTIONAL RESOURCES

- (20 min.) Have students do **Cuaderno del escritor,** p. 153. See **Cuaderno del escritor,** *TRB,* p. 84.
- (20 min.) Do **Investigación y presentación,** p. 153. See **Investigación y presentación,** *TRB,* p. 84.
- (20 min.) Do **Crear modelos,** p. 153. See **Crear modelos,** *TRB,* p. 85.
- (20 min.) Do **Escribir cartas,** p. 153. See **Escribir cartas,** *TRB,* p. 85.

Practice Options/Homework Suggestions

- Internet (go.hrw.com, Keyword: WN3 AMBIENTE-LEC)
- Have students answer questions in **Más allá del texto,** p. 152.
- Have students study for **Prueba de lectura.**

▲ = Advanced Learners ◆ = Slower Pace Learners ● = Special Learning Needs ■ = Heritage Speakers

Teacher's Name ______________ Class ______________ Date ______________

COLECCIÓN 3

El frágil medio ambiente

DAY 4 50-MINUTE LESSON PLAN

STANDARDS FOR FOREIGN LANGUAGE LEARNING: DAY 4

Lectura/Gramática

Communication 1.2: Students understand and interpret written and spoken language on a variety of topics.

Comparisons 4.1: Students demonstrate understanding of the nature of language through comparisons of the language studied and their own.

CORE INSTRUCTION

Warm-Up

- (5 min.) Have students review **"La fiesta del árbol,"** pp. 149–151.

Lectura

Assess

- (30 min.) Give **Prueba de lectura: "La fiesta del árbol,"** *Assessment Program,* pp. 79–80.

Gramática

Teach

- (10 min.) Present **Los usos de *se*,** pp. 194–195. See **Los usos de *se*,** *TRB,* p. 104.

Wrap-Up

- (5 min.) Have students do Additional Practice, *TRB,* p. 104.

OPTIONAL RESOURCES

- (10 min.) Have students do **Para hispanohablantes,** *TRB,* p. 107. ■
- (15 min.) Have students do **Para angloparlantes,** *TRB,* p. 107.

Practice Options/Homework Suggestions

- *Cuaderno de práctica,* Activities 1–4, pp. 56–57
- *Advanced Placement Literature Preparation Book,* pp. 51–52 ▲

▲ = Advanced Learners ◆ = Slower Pace Learners ● = Special Learning Needs ■ = Heritage Speakers

Teacher's Name ______________ Class ______________ Date ______________

COLECCIÓN **3**

El frágil medio ambiente

DAY 5 50-MINUTE LESSON PLAN

STANDARDS FOR FOREIGN LANGUAGE LEARNING: DAY 5

Gramática

Communication 1.2: Students understand and interpret written and spoken language on a variety of topics.

Comparisons 4.1: Students demonstrate understanding of the nature of language through comparisons of the language studied and their own.

CORE INSTRUCTION

Warm-Up

- (5 min.) Have students review **Los usos de *se*,** pp. 194–195.

Gramática

Teach

- (10 min.) Have students do Activity A, p. 195.
- (10 min.) Have students do Activity B, p. 196.
- (10 min.) Have students do Activity D, p. 196.
- (10 min.) Have students do Activity E, p. 197.

Wrap-Up

- (5 min.) Have students answer questions in Activity F, p. 197. See Activity F, *TRB*, p. 105.

OPTIONAL RESOURCES

- (15 min.) Have students do **Ampliación: Hoja de práctica 3-A, Más sobre *se* como pronombre de complemento indirecto,** *TRB*, p. 283. ◆ ●
- (20 min.) Have students do Activity C, p. 196.
- (20 min.) Have students do Activity G, p. 198.
- (20 min.) Have students do Activity H, p. 198.
- (20 min.) Have students do Pair Work, *TRB*, p. 105.

Practice Options/Homework Suggestions

- *Cuaderno de práctica,* Activities 5–10, pp. 58–60
- *Cuaderno de práctica,* **Ampliación, Hoja de práctica 3-A,** p. 142 ◆ ●
- Have students study for **Prueba de gramática.**

▲ = Advanced Learners ◆ = Slower Pace Learners ● = Special Learning Needs ■ = Heritage Speakers

El frágil medio ambiente

DAY 6 50-MINUTE LESSON PLAN

STANDARDS FOR FOREIGN LANGUAGE LEARNING: DAY 6

Gramática/Cultura y lengua

Communication 1.1: Students engage in conversations, provide and obtain information, express feelings and emotions, and exchange opinions.

Communication 1.2: Students understand and interpret written and spoken language on a variety of topics.

Cultures 2.1: Students demonstrate an understanding of the relationship between the practices and perspectives of the culture studied.

Cultures 2.2: Students demonstrate an understanding of the relationship between the products and perspectives of the culture studied.

Connections 3.1: Students reinforce and further their knowledge of other disciplines through the foreign language.

Connections 3.2: Students acquire information and recognize the distinctive viewpoints that are only available through the foreign language and its cultures.

CORE INSTRUCTION

Warm-Up

- (5 min.) Have students review **Los usos de *se*,** pp. 194–198.

Gramática

Assess

- (20 min.) Give **Prueba de gramática: Los usos de *se*,** *Assessment Program,* p. 89.

Cultura y lengua

Teach

- (10 min.) Have students read aloud **Cultura y lengua,** pp. 164–165.
- (10 min.) Have groups of students do Thinking Critically, *TRB,* pp. 90–91.

Wrap-Up

- (5 min.) Have students discuss the map and photographs of **Chile,** pp. 164–165.

OPTIONAL RESOURCES

- (15 min.) See **Cultura y lengua,** Teaching Suggestions, *Video Guide,* p. 21.
- (20 min.) Have students do History Link, *TRB,* p. 91.

Practice Options/Homework Suggestions

- Internet (go.hrw.com, Keyword: WN3 AMBIENTE-CYL)
- Have students read **Cultura y lengua,** pp. 166–167.
- *Advanced Placement Literature Preparation Book,* pp. 53–56 ▲

Assessment Options

- *Assessment Program,* **Prueba de gramática 3-A: Más sobre *se* come pronombre de complemento indirecto,** p. 261

▲ = Advanced Learners ◆ = Slower Pace Learners ● = Special Learning Needs ■ = Heritage Speakers

El frágil medio ambiente

DAY 7 50-MINUTE LESSON PLAN

STANDARDS FOR FOREIGN LANGUAGE LEARNING: DAY 7

Cultura y lengua

Communication 1.1: Students engage in conversations, provide and obtain information, express feelings and emotions, and exchange opinions.

Communication 1.3: Students present information, concepts, and ideas to an audience of listeners or readers on a variety of topics.

Cultures 2.2: Students demonstrate an understanding of the relationship between the products and perspectives of the culture studied.

Connections 3.1: Students reinforce and further their knowledge of other disciplines through the foreign language.

Comparisons 4.1: Students demonstrate understanding of the nature of language through comparisons of the language studied and their own.

CORE INSTRUCTION

Warm-Up

- (5 min.) Have students discuss Getting Started, *TRB,* p. 90.

Cultura y lengua

Teach

- (5 min.) Present **Así se dice,** p. 168. See **Así se dice,** *TRB,* p. 91.
- (15 min.) Have pairs of students do **Actividad,** p. 168.
- (10 min.) Show **Cultura y lengua: Chile,** *Video Program* (Videocassette 1). See Teaching Suggestions, **Mientras lo ves,** *Video Guide,* p. 21.
- (5 min.) Do **Hoja de actividades 2, Mientras lo ves,** *Video Guide,* p. 25.

Wrap-Up

- (10 min.) Discuss **Modismos y regionalismos,** p. 168. See Language Note, *TRB,* p. 91.

OPTIONAL RESOURCES

- (20 min.) Have students do Science Link, *TRB,* p. 90 ▲ ■
- (10 min.) Have students do **Hoja de actividades 2,** *Video Guide,* p. 25.
- (30 min.) Do Teaching Suggestions, **Después de ver,** *Video Guide,* p. 21.

Practice Options/Homework Suggestions

- Internet (go.hrw.com, Keyword: WN3 AMBIENTE-CYL)
- Have students study for **Prueba de cultura.**

▲ = Advanced Learners ◆ = Slower Pace Learners ● = Special Learning Needs ■ = Heritage Speakers

El frágil medio ambiente

DAY 8 50-MINUTE LESSON PLAN

STANDARDS FOR FOREIGN LANGUAGE LEARNING: DAY 8

Cultura y lengua/Elementos de literatura

Communication 1.2: Students understand and interpret written and spoken language on a variety of topics.

Connections 3.1: Students reinforce and further their knowledge of other disciplines through the foreign language.

Communities 5.2: Students show evidence of becoming life-long learners by using the language for personal enjoyment and enrichment.

CORE INSTRUCTION

Warm-Up

- (5 min.) Have students review **Cultura y lengua,** pp. 164–168.

Cultura y lengua

Assess

- (20 min.) Give **Prueba de cultura: Chile,** *Assessment Program,* p. 93.

Elementos de literatura

Teach

- (20 min.) Have students read **Elementos de literatura: Rima, Ritmo,** and **Repetición y paralelismo,** pp. 169–170.

Wrap-Up

- (5 min.) See Kinesthetic Link, *TRB,* pp. 92–93.

OPTIONAL RESOURCES

- (20 min.) Have students do Music Link, *TRB,* p. 93.

Practice Options/Homework Suggestions

- Internet (go.hrw.com, Keyword: WN3 AMBIENTE-CYL)
- Have students read **Elementos de literatura,** pp. 170–172. ▲

▲ = Advanced Learners ◆ = Slower Pace Learners ● = Special Learning Needs ■ = Heritage Speakers

El frágil medio ambiente

DAY 9 50-MINUTE LESSON PLAN

STANDARDS FOR FOREIGN LANGUAGE LEARNING: DAY 9

Elementos de literatura/Lectura

Communication 1.2: Students understand and interpret written and spoken language on a variety of topics.

Connections 3.1: Students reinforce and further their knowledge of other disciplines through the foreign language.

Comparisons 4.1: Students demonstrate understanding of the nature of language through comparisons of the language studied and their own.

CORE INSTRUCTION

Warm-Up

- (5 min.) Have students identify as many elements of poetry as they can in the poems shown in Getting Started, *TRB,* p. 92.

Elementos de literatura

Teach

- (10 min.) Have students do the first Additional Practice, *TRB,* p. 93.
- (5 min.) Have students do the second Additional Practice, *TRB,* p. 93.

Lectura

Teach

- (25 min.) Have volunteers read aloud **"Árbol adentro," "Paisaje,"** and **"Meciendo,"** pp. 157–159. See **Elementos de literatura** and Techniques for Handling the Reading, *TRB,* pp. 87–88, and Applying the Element, *TRB,* p. 92.

Wrap-Up

- (5 min.) Have students do **Comparte tus ideas,** p. 156. See **Comparte tus ideas,** *TRB,* p. 87.

OPTIONAL RESOURCES

- (5 min.) Read aloud **"Árbol adentro," "Paisaje,"** and **"Meciendo,"** Summary, *TRB,* p. 87. ◆ ●
- (10 min.) Have students listen to the poems on Audio CD 3, Tracks 5–7.

Practice Options/Homework Suggestions

- Internet (go.hrw.com, Keyword: WN3 AMBIENTE-LEC)
- Have students study **Vocabulario esencial, "Árbol adentro," "Paisaje," "Meciendo,"** p. 215. ◆ ●
- Have students study **Vocabulario adicional, "Árbol adentro," "Paisaje" "Meciendo,"** *TRB,* p. 299. ◆ ●
- Have students read **Conoce a los escritores,** pp. 160–161.
- *Advanced Placement Literature Preparation Book,* pp. 57–62 ▲

▲ = Advanced Learners ◆ = Slower Pace Learners ● = Special Learning Needs ■ = Heritage Speakers

Teacher's Name ______________________ Class ____________________ Date __________

COLECCIÓN 3

El frágil medio ambiente

DAY 10 50-MINUTE LESSON PLAN

STANDARDS FOR FOREIGN LANGUAGE LEARNING: DAY 10

Lectura

Communication 1.2: Students understand and interpret written and spoken language on a variety of topics.

Communication 1.3: Students present information, concepts, and ideas to an audience of listeners or readers on a variety of topics.

Communities 5.1: Students use the language both within and beyond the school setting.

Communities 5.2: Students show evidence of becoming life-long learners by using the language for personal enjoyment and enrichment.

CORE INSTRUCTION

Warm-Up

- (5 min.) Have students do **Vocabulario en contexto,** Activity C, pp. 189–190.

Lectura

Teach

- (5 min.) Have students do **Primeras impresiones,** p. 162. See **Primeras impresiones,** *TRB,* p. 88.
- (10 min.) Have students do **Interpretaciones del texto,** p. 162. See **Interpretaciones del texto,** *TRB,* p. 88.
- (5 min.) Present **Así se dice,** p. 162.
- (20 min.) Have students do **Cuaderno del escritor** using the expressions from **Así se dice,** p. 162. See **Cuaderno del escritor,** *TRB,* p. 89.

Wrap-Up

- (5 min.) Have students do **Más allá del texto,** p. 162.

OPTIONAL RESOURCES

- (30 min.) Have students practice the Reading Strategy, *TRB,* pp. 255–257. ▲ ◆ ●
- (20 min.) Have students do **Para hispanohablantes,** *TRB,* p. 89. ■
- (20 min.) Have students do **Para angloparlantes,** *TRB,* p. 89.
- (30 min.) Have students do **Publicación y poesía,** p. 163. See **Publicación y poesía,** *TRB,* p. 89. ▲
- (30 min.) Have students do **Arte,** p. 162. See **Arte,** *TRB,* p. 89. ◆ ●

Practice Options/Homework Suggestions

- Internet (go.hrw.com, Keyword: WN3 AMBIENTE-LEC)
- *Cuaderno de práctica,* Activities 1–3, pp. 47–48
- Have students study for **Prueba de lectura.**

▲ = Advanced Learners ◆ = Slower Pace Learners ● = Special Learning Needs ■ = Heritage Speakers

Teacher's Name ______________________ Class ______________________ Date ____________

COLECCIÓN 3

El frágil medio ambiente

DAY 11 50-MINUTE LESSON PLAN

STANDARDS FOR FOREIGN LANGUAGE LEARNING: DAY 11

Lectura

Communication 1.1: Students engage in conversations, provide and obtain information, express feelings and emotions, and exchange opinions.

Communication 1.2: Students understand and interpret written and spoken language on a variety of topics.

Communication 1.3: Students present information, concepts, and ideas to an audience of listeners or readers on a variety of topics.

Connections 3.1: Students reinforce and further their knowledge of other disciplines through the foreign language.

CORE INSTRUCTION

Warm-Up

- (5 min.) Have students review **"Árbol adentro," "Paisaje,"** and **"Meciendo,"** pp. 157–159.

Lectura
Assess

- (20 min.) Give **Prueba de lectura: "Árbol adentro," "Paisaje," y "Meciendo,"** *Assessment Program,* pp. 81–82.

Lectura
Teach

- (10 min.) Have students read **Estrategias para leer,** p. 154. See Introducing the Strategy, *TRB,* p. 86.
- (10 min.) See Getting Started, *TRB,* p. 86.

Wrap-Up

- (5 min.) Have students do the Group Work activity, *TRB,* p. 86.

OPTIONAL RESOURCES

- (15 min.) Have students do Extension, *TRB,* p. 86. ◆ ●

Practice Options/Homework Suggestions

- Internet (go.hrw.com, Keyword: WN3 AMBIENTE-LEC)
- *Advanced Placement Literature Preparation Book,* pp. 63–66 ▲

▲ = Advanced Learners ◆ = Slower Pace Learners ● = Special Learning Needs ■ = Heritage Speakers

El frágil medio ambiente

DAY 12 50-MINUTE LESSON PLAN

STANDARDS FOR FOREIGN LANGUAGE LEARNING: DAY 12

Lectura

Communication 1.2: Students understand and interpret written and spoken language on a variety of topics.

Connections 3.1: Students reinforce and further their knowledge of other disciplines through the foreign language.

CORE INSTRUCTION

Warm-Up

- (5 min.) Have students read **Punto de partida** and **Elementos de literatura,** p. 173.

Lectura

Teach

- (15 min.) Have students do **Punto de partida,** *TRB,* p. 94.
- (10 min.) Have students do **Elementos de literatura,** *TRB,* p. 94.
- (15 min.) Have students listen to the audio recording of the first part of **"Las abejas de bronce";** you may wish to use Audio CD 3, Track 8.

Wrap-Up

- (5 min.) Have students do **Primeras impresiones,** p. 182.

OPTIONAL RESOURCES

- (10 min.) Read aloud **"Las abejas de bronce,"** Summary, *TRB,* p. 94. ◆ ●
- (10 min.) Have students do Techniques for Handling the Reading, *TRB,* pp. 94–95. ◆ ●

Practice Options/Homework Suggestions

- Internet (go.hrw.com, Keyword: WN3 AMBIENTE-LEC)
- Have students finish reading **"Las abejas de bronce,"** pp. 174–181.
- Have students study **Vocabulario esencial, "Las abejas de bronce,"** p. 215. ◆ ●
- *Cuaderno de práctica,* Activities 1–2, p. 49

▲ = Advanced Learners ◆ = Slower Pace Learners ● = Special Learning Needs ■ = Heritage Speakers

El frágil medio ambiente

DAY 13 50-MINUTE LESSON PLAN

STANDARDS FOR FOREIGN LANGUAGE LEARNING: DAY 13

Lectura

Communication 1.2: Students understand and interpret written and spoken language on a variety of topics.

Communication 1.3: Students present information, concepts, and ideas to an audience of listeners or readers on a variety of topics.

Connections 3.1: Students reinforce and further their knowledge of other disciplines through the foreign language.

Communities 5.1: Students use the language both within and beyond the school setting.

Communities 5.2: Students show evidence of becoming life-long learners by using the language for personal enjoyment and enrichment.

CORE INSTRUCTION

Warm-Up

- (5 min.) Have students read **Conoce al escritor,** p. 181.

Lectura

Teach

- (10 min.) Have students listen to the rest of the audio recording of **"Las abejas de bronce,"** Audio CD 3, Track 8.
- (15 min.) Have students do **Repaso del texto,** p. 182.
- (15 min.) Have students do **Interpretaciones del texto,** p. 182.

Wrap-Up

- (5 min.) Have students do **Preguntas al texto,** p. 182.

OPTIONAL RESOURCES

- (10 min.) Have students do **Para hispanohablantes,** *TRB,* p. 96. ■
- (10 min.) Have students do **Para angloparlantes,** *TRB,* p. 96.

Practice Options/Homework Suggestions

- Internet (go.hrw.com, Keyword: WN3 AMBIENTE-LEC)
- Have students practice the Reading Strategy, *TRB,* pp. 258–260.
- Have students study **Vocabulario adicional, "Las abejas de bronce,"** *TRB,* p. 299. ◆ ●
- *Cuaderno de práctica,* Activities 3–4, p. 50
- *Cuaderno de práctica,* **Vocabulario adicional,** p. 141

▲ = Advanced Learners ◆ = Slower Pace Learners ● = Special Learning Needs ■ = Heritage Speakers

Teacher's Name ______________________ Class ____________________ Date __________

COLECCIÓN 3

El frágil medio ambiente

DAY 14 50-MINUTE LESSON PLAN

STANDARDS FOR FOREIGN LANGUAGE LEARNING: DAY 14

Lectura/Vocabulario

Communication 1.2: Students understand and interpret written and spoken language on a variety of topics.

Communication 1.3: Students present information, concepts, and ideas to an audience of listeners or readers on a variety of topics.

Connections 3.1: Students reinforce and further their knowledge of other disciplines through the foreign language.

Communities 5.1: Students use the language both within and beyond the school setting.

CORE INSTRUCTION

Warm-Up

- (5 min.) Have students study **Así se dice,** p. 182.

Lectura

Teach

- (5 min.) Present **Así se dice,** p. 182. See **Así se dice,** *TRB,* p. 95.
- (15 min.) Have students do **Escritura,** p. 183. See **Escritura,** *TRB,* p. 96.

Vocabulario

Teach

- (10 min.) Have students do Activity D, p. 190. See Activity D, *TRB,* p. 101.
- (10 min.) Have students do Activity E, p. 190. You may wish to use Audio CD 1, Track 20. See script, *TRB,* p. 102.

Wrap-Up

- (5 min.) Have volunteers read their **fábulas** to the class.

OPTIONAL RESOURCES

- (30 min.) Have students do **Dramatización,** p. 183. ▲
- (30 min.) Have students do **Escribir un panfleto informativo,** p. 183. ▲
- (20 min.) Have students do the first Extension, *TRB,* p. 102. ◆ ●

Practice Options/Homework Suggestions

- Internet (go.hrw.com, Keyword: WN3 AMBIENTE-LEC)
- Have students do **Cuaderno del escritor,** p. 183. ▲
- Have students study for **Prueba de lectura.**
- *Advanced Placement Literature Preparation Book,* pp. 67–69 ▲

▲ = Advanced Learners ◆ = Slower Pace Learners ● = Special Learning Needs ■ = Heritage Speakers

El frágil medio ambiente

STANDARDS FOR FOREIGN LANGUAGE LEARNING: DAY 15

Lectura/Vocabulario

Communication 1.2: Students understand and interpret written and spoken language on a variety of topics.

Communication 1.3: Students present information, concepts, and ideas to an audience of listeners or readers on a variety of topics.

Connections 3.1: Students reinforce and further their knowledge of other disciplines through the foreign language.

Communities 5.1: Students use the language both within and beyond the school setting.

Communities 5.2: Students show evidence of becoming life-long learners by using the language for personal enjoyment and enrichment.

CORE INSTRUCTION

Warm-Up

- (5 min.) Have student review **"Las abejas de bronce,"** pp. 174–181.

Lectura

Assess

- (20 min.) Give **Prueba de lectura: "Las abejas de bronce,"** pp. 83–84.

Vocabulario

Teach

- (10 min.) Present **Mejora tu vocabulario: El lenguaje figurado,** p. 191. See **El lenguaje figurado,** *TRB,* p. 102.
- (10 min.) Have students do Activity F, p. 191.

Wrap-Up

- (5 min.) Have student do the second Extension, *TRB,* p. 102.

OPTIONAL RESOURCES

- (15 min.) Have students read **A leer por tu cuenta: "Dicen que no hablan las plantas,"** p. 186. ▲
- (30 min.) See Summary, Background Information, **Antes de leer,** and Techniques for Handling the Reading, *TRB,* pp. 98–99. ▲
- (30 min.) Have students select from Additional Practice, Extension, Music Link, and Art Link, *TRB,* p. 99. ▲
- (10 min.) Have students do **Para hispanohablantes,** *TRB,* p. 99. ■
- (10 min.) Have students do **Para angloparlantes,** *TRB,* p. 99.
- (30 min.) Have students do **"Dicen que no hablan las plantas,"** *Cuaderno de práctica,* pp. 51–53. ▲
- (20 min.) Have students practice the Reading Strategy, *TRB,* pp. 261–262. ◆ ●

Practice Options/Homework Suggestions

- Have students study **Vocabulario esencial, Mejora tu vocabulario,** p. 215. ◆ ●
- Have students do Activity I or J, p. 193.
- Have students study for **Prueba de vocabulario.**
- *Cuaderno de práctica,* Activities 1–4, pp. 54–55

Assessment Options

- *Assessment Program,* **Prueba de vocabulario adicional,** p. 260

▲ = Advanced Learners ◆ = Slower Pace Learners ● = Special Learning Needs ■ = Heritage Speakers

Teacher's Name ______________________ Class ______________________ Date ____________

COLECCIÓN 3

El frágil medio ambiente

DAY 16 50-MINUTE LESSON PLAN

STANDARDS FOR FOREIGN LANGUAGE LEARNING: DAY 16

Vocabulario/Gramática

Communication 1.2: Students understand and interpret written and spoken language on a variety of topics.

Comparisons 4.1: Students demonstrate understanding of the nature of language through comparisons of the language studied and their own.

Communities 5.1: Students use the language both within and beyond the school setting.

CORE INSTRUCTION

Warm-Up

- (5 min.) Have students review **El lenguaje figurado,** p. 191.

Vocabulario

Teach

- (5 min.) Have students do Activity G, p. 192.
- (5 min.) As a review for the quiz, have students do Additional Practice, *TRB,* p. 103.

Vocabulario

Assess

- (20 min.) Give **Prueba de vocabulario,** *Assessment Program,* pp. 87–88.

Gramática

Teach

- (10 min.) Present **La voz pasiva: Formas de la voz pasiva,** pp. 198–199. See **Formas de la voz pasiva,** *TRB,* pp. 105–106.

Wrap-Up

- (5 min.) Present **Se usa la voz pasiva con *ser*...,** p. 199. See **Se usa la voz pasiva con *ser*...,** *TRB,* p. 106.

OPTIONAL RESOURCES

- (15 min.) Have students do Additional Practice, *TRB,* p. 106. ◆ ●
- (10 min.) Have students do Activity H, p. 192.
- (10 min.) Have students do **Para hispanohablantes,** *TRB,* p. 103. ■
- (10 min.) Have students do **Para angloparlantes,** *TRB,* p. 103.

Practice Options/Homework Suggestions

- *Cuaderno de práctica,* Activities 11–14, pp. 61–62
- *Advanced Placement Literature Preparation Book,* pp. 70–72 ▲

Assessment Options

- *Assessment Program,* **Prueba de lectura: "Dicen que no hablan las plantas,"** pp. 85–86

▲ = Advanced Learners ◆ = Slower Pace Learners ● = Special Learning Needs ■ = Heritage Speakers

El frágil medio ambiente

DAY 17 50-MINUTE LESSON PLAN

STANDARDS FOR FOREIGN LANGUAGE LEARNING: DAY 17

Gramática

Communication 1.2: Students understand and interpret written and spoken language on a variety of topics.

Comparisons 4.1: Students demonstrate understanding of the nature of language through comparisons of the language studied and their own.

CORE INSTRUCTION

Warm-Up

- (5 min.) Have students review **La voz pasiva,** pp. 198–199.

Gramática

Teach

- (10 min.) Have students do Activity I, p. 200.
- (10 min.) Have students do Activity J, items 1–5, p. 200.
- (10 min.) Have students do Activity K, items 1–5, p. 200.
- (10 min.) Have students do Activity L, p. 201.

Wrap-Up

- (5 min.) Review the rules of **la voz pasiva.**

OPTIONAL RESOURCES

- (10 min.) Have students do Activity J, items 6–10, p. 200.
- (10 min.) Have students do Activity K, items 6–10, p. 200.

Practice Options/Homework Suggestions

- Have students do Activity M or N, p. 201.
- Have students study for **Prueba de gramática.**

▲ = Advanced Learners ◆ = Slower Pace Learners ● = Special Learning Needs ■ = Heritage Speakers

Teacher's Name ______________________ Class ______________________ Date ____________

COLECCIÓN 3

El frágil medio ambiente

DAY 18 50-MINUTE LESSON PLAN

STANDARDS FOR FOREIGN LANGUAGE LEARNING: DAY 18

Gramática

Communication 1.2: Students understand and interpret written and spoken language on a variety of topics.

Comparisons 4.1: Students demonstrate understanding of the nature of language through comparisons of the language studied and their own.

CORE INSTRUCTION

Warm-Up

- (5 min.) Have students review **"La voz pasiva,"** pp. 198–201.

Gramática

Assess

- (30 min.) Give **Prueba de gramática: La voz pasiva,** *Assessment Program,* pp. 90–91.

Gramática

Teach

- (5 min.) Present **Comparación y contraste,** p. 202.
- (5 min.) Have students do Activity A, items 1–5, p. 203.

Wrap-Up

- (5 min.) Have students do Activity C, items 1–5, p. 203.

OPTIONAL RESOURCES

- (15 min.) Have students do Additional Practice, *TRB,* p. 106. ◆ ●
- (5 min.) Have students do Activity A, items 6–10, p. 203.
- (5 min.) Have students do Activity C, items 6–11, p. 203.

Practice Options/Homework Suggestions

- Have students read **Panorama cultural,** pp. 184–185.
- Have students do Activity B, p. 203.
- *Cuaderno de práctica,* Activities 15–16, p. 63
- Have students study for **Prueba de comparación y contraste.**

▲ = Advanced Learners ◆ = Slower Pace Learners ● = Special Learning Needs ■ = Heritage Speakers

Teacher's Name ______________________ Class ____________________ Date __________

COLECCIÓN 3

El frágil medio ambiente

DAY 19 50-MINUTE LESSON PLAN

STANDARDS FOR FOREIGN LANGUAGE LEARNING: DAY 19

Gramática/Panorama cultural

Communication 1.2: Students understand and interpret written and spoken language on a variety of topics.

Communication 1.3: Students present information, concepts, and ideas to an audience of listeners or readers on a variety of topics.

Cultures 2.1: Students demonstrate an understanding of the relationship between the practices and perspectives of the culture studied.

Cultures 2.2: Students demonstrate an understanding of the relationship between the products and perspectives of the culture studied.

Connections 3.2: Students acquire information and recognize the distinctive viewpoints that are only available through the foreign language and its cultures.

Comparisons 4.2: Students demonstrate understanding of the concept of culture through comparisons of the cultures studied and their own.

CORE INSTRUCTION

Warm-Up

- (5 min.) Have students review **Comparación y contraste,** pp. 202–203.

Gramática

Assess

- (20 min.) Give **Prueba de comparación y contraste,** *Assessment Program,* p. 92.

Panorama cultural

Teach

- (5 min.) Have students discuss the introduction of **Panorama cultural,** p. 184. See Presentation, **Panorama cultural,** *TRB,* p. 97.
- (15 min.) Show **Panorama cultural,** Video Program (Videocassette 1). See **Panorama cultural,** Teaching Suggestions, **Mientras lo ves,** *Video Guide,* p. 22.

Wrap-Up

- (5 min.) Have students do **Hoja de actividades 3, Mientras lo ves,** *Video Guide,* p. 26.

OPTIONAL RESOURCES

- (5 min.) Play Audio CD 1, Tracks 17–19, and have students listen to the interviews.
- (20 min.) Have students do **Panorama cultural,** Teaching Suggestions, **Antes de ver,** *Video Guide,* p. 22.
- (20 min.) Have students do **Panorama cultural,** Teaching Suggestions, **Después de ver,** *Video Guide,* p. 22.
- (10 min.) Have students do Activity D, p. 185. You may use Audio CD 1, Track 19. See script, *TRB,* p. 97.
- (20 min.) Have students do Thinking Critically, *TRB,* p. 97. ▲

Practice Options/Homework Suggestions

- Have students read **Comunidad y oficio,** p. 188.
- Have students write answers to **Para pensar y hablar,** Activities A and B, p. 185.
- *Advanced Placement Literature Preparation Book,* pp. 73–75 ▲

▲ = Advanced Learners ◆ = Slower Pace Learners ● = Special Learning Needs ■ = Heritage Speakers

El frágil medio ambiente

DAY 20 50-MINUTE LESSON PLAN

STANDARDS FOR FOREIGN LANGUAGE LEARNING: DAY 20

Comunidad y oficio

Communication 1.1: Students engage in conversations, provide and obtain information, express feelings and emotions, and exchange opinions.

Cultures 2.2: Students demonstrate an understanding of the relationship between the products and perspectives of the culture studied.

Connections 3.1: Students reinforce and further their knowledge of other disciplines through the foreign language.

Comparisons 4.2: Students demonstrate understanding of the concept of culture through comparisons of the cultures studied and their own.

CORE INSTRUCTION

Warm-Up

- (5 min.) Have pairs of students do Activity C, **Panorama cultural,** p. 185.

Comunidad y oficio

Teach

- (5 min.) Present **Comunidad y oficio,** p. 188. See Getting Started, *TRB,* p.100.
- (10 min.) Have students read and discuss **Comunidad y oficio,** p. 188.
- (10 min.) Have students do **Comunidad y oficio,** Teaching Suggestions, **Antes de ver,** *Video Guide,* p. 23.
- (10 min.) Show **Comunidad y oficio,** *Video Program* (Videocassette 1). See **Comunidad y oficio,** Teaching Suggestions, **Mientras lo ves,** *Video Guide,* p. 23.
- (5 min.) Have students do **Hoja de actividades 4, Mientras lo ves,** *Video Guide,* p. 27.

Wrap-Up

- (5 min.) Have students do **Hoja de actividades 4, Después de ver,** *Video Guide,* p. 27.

OPTIONAL RESOURCES

- (10 min.) Review Background Information, *TRB,* p. 100.
- (30 min.) Have students do Activity A, p. 188. See Community Link, *TRB,* p. 100. ▲
- (30 min.) Have students do Activity B, p. 188.
- (30 min.) Have students do Thinking Critically, *TRB,* p. 100. ▲

Practice Options/Homework Suggestions

- Internet (go.hrw.com, Keyword: WN3 AMBIENTE-CYO)

▲ = Advanced Learners ◆ = Slower Pace Learners ● = Special Learning Needs ■ = Heritage Speakers

Teacher's Name ______________________ Class ______________________ Date ______________

COLECCIÓN 3

El frágil medio ambiente

DAY 21 50-MINUTE LESSON PLAN

STANDARDS FOR FOREIGN LANGUAGE LEARNING: DAY 21

Ortografía

Communication 1.2: Students understand and interpret written and spoken language on a variety of topics.

Comparisons 4.1: Students demonstrate understanding of the nature of language through comparisons of the language studied and their own.

CORE INSTRUCTION

Warm-Up

- (5 min.) Have students read the introduction of **Los sonidos /b/ y /v/,** p. 204.

Ortografía

Teach

- (15 min.) Present **Los sonidos /b/ y /v/,** p. 204. See **Los sonidos /b/ y /v/** and Language Note, *TRB,* p. 108.
- (10 min.) Have students do Activities A and B, p. 205.
- (5 min.) Have students do Activity C, p. 206.
- (10 min.) Present **La acentuación,** pp. 206–207. See **La acentuación,** *TRB,* p.109.

Wrap-Up

- (5 min.) Do Activity D, p. 207, as a class.

OPTIONAL RESOURCES

- (10 min.) Have students do Pair Work, *TRB,* p. 108. ◆ ●
- (10 min.) Have students do Additional Practice, *TRB,* p. 108. ◆ ●
- (20 min.) Have students do Cooperative Learning, *TRB,* p. 108. ◆ ●
- (15 min.) Have students do **Ampliación, Hoja de práctica 3-B, Las letras *b, d* y *g* intervocálicas,** *TRB,* p. 284. ◆ ●
- (10 min.) Have students do **Ampliación, Hoja de práctica 3-C, Confusión entre /p/ o /b/ y /k/ o /g/,** *TRB,* p. 285. ◆ ●

Practice Options/Homework Suggestions

- *Cuaderno de práctica,* Activities 1–3, p. 64
- *Cuaderno de práctica,* **Ampliación, Hoja de práctica 3-B,** p. 143
- *Cuaderno de práctica,* **Ampliación, Hoja de práctica 3-C,** p. 144
- *Advanced Placement Literature Preparation Book,* pp. 76–79 ▲

▲ = Advanced Learners ◆ = Slower Pace Learners ● = Special Learning Needs ■ = Heritage Speakers

Teacher's Name ______________________ Class ______________________ Date ____________

COLECCIÓN 3

El frágil medio ambiente

DAY 22 50-MINUTE LESSON PLAN

STANDARDS FOR FOREIGN LANGUAGE LEARNING: DAY 22

Ortografía

Communication 1.2: Students understand and interpret written and spoken language on a variety of topics.

Comparisons 4.1: Students demonstrate understanding of the nature of language through comparisons of the language studied and their own.

CORE INSTRUCTION

Warm-Up

- (5 min.) Have students review **Ortografía,** pp. 204–206.

Ortografía

Teach

- (10 min.) Have students do Activity E, p. 207. See Activity E, *TRB,* p. 109.
- (10 min.) Have students do Activity F, p. 207. See Activity F, *TRB,* p. 109.
- (20 min.) Give **Dictado,** Activities A and B, p. 207. You may wish to use Audio CD 1, Tracks 21 and 22. See scripts, *TRB,* p. 110.

Wrap-Up

- (5 min.) Review spelling rules for ***/b/*** and ***/v/,*** and **diptongos** and **hiatos.**

OPTIONAL RESOURCES

- (15 min.) Have students do Challenge, *TRB,* p. 109. ▲
- (20 min.) Have students do **Para hispanohablantes,** *TRB,* p. 110. ■
- (20 min.) Have students do **Para angloparlantes,** *TRB,* p. 110.
- (15 min.) Have students do **Ampliación, Hoja de práctica 3-D, Metátesis de diptongos,** *TRB,* p. 286.

Practice Options/Homework Suggestions

- *Cuaderno de práctica,* Activities 4–6, p. 65
- *Cuaderno de práctica,* **Ampliación, Hoja de práctica 3-D,** p. 145
- Have students study for **Prueba de ortografía.**

Assessment Options

- *Assessment Program,* **Prueba de ortografía 3-B: Las letras *b, d* y *g* intervocales,** p. 262
- *Assessment Program,* **Prueba de ortografía 3-C: Confusión entre los sonidos /p/ o /b/ y /k/ o /g/,** p. 263

▲ = Advanced Learners ◆ = Slower Pace Learners ● = Special Learning Needs ■ = Heritage Speakers

Teacher's Name ______________________ Class ______________________ Date __________

COLECCIÓN 3

El frágil medio ambiente

DAY 23 50-MINUTE LESSON PLAN

STANDARDS FOR FOREIGN LANGUAGE LEARNING: DAY 23

Ortografía/Taller del escritor

Communication 1.2: Students understand and interpret written and spoken language on a variety of topics.

Connections 3.1: Students reinforce and further their knowledge of other disciplines through the foreign language.

Communities 5.1: Students use the language both within and beyond the school setting.

Communities 5.2: Students show evidence of becoming life-long learners by using the language for personal enjoyment and enrichment.

CORE INSTRUCTION

Warm-Up

- (5 min.) Have students review **Ortografía,** pp. 204–207.

Ortografía

Assess

- (10 min.) Give **Prueba de ortografía,** *Assessment Program,* p. 94.

Taller del escritor

Teach

- (5 min.) Have students review their work on the portfolio suggestions after each reading selection. See **Opciones: Prepara tu portafolio,** pp. 153, 163 and 183.
- (10 min.) Introduce **Taller del escritor,** p. 208. Have students read **La exposición: Artículo informativo,** p. 208.
- (15 min.) Introduce **Antes de escribir: Preguntas** and **Piensa en los lectores y en la idea principal,** pp. 208–209. See **Antes de escribir,** *TRB,* p. 111.

Wrap-Up

- (5 min.) Review **Instrucciones para escoger un tema** with students, p. 208.

OPTIONAL RESOURCES

- (10 min.) See Presenting the Workshop, *TRB,* p. 111.
- (10 min.) Have students do **Antes de escribir,** *TRB,* p. 111.

Practice Options/Homework Suggestions

- Have students choose the topic of their article. See **Tarea,** p. 208.

Assessment Options

- *Assessment Program,* **Prueba de ortografía 3-D: Metátesis de diptongos,** p. 264

▲ = Advanced Learners ◆ = Slower Pace Learners ● = Special Learning Needs ■ = Heritage Speakers

Teacher's Name ______________________ Class ______________________ Date ______________

COLECCIÓN 3

El frágil medio ambiente

DAY 24 50-MINUTE LESSON PLAN

STANDARDS FOR FOREIGN LANGUAGE LEARNING: DAY 24

Taller del escritor

Connections 3.1: Students reinforce and further their knowledge of other disciplines through the foreign language.

Comparisons 4.1: Students demonstrate understanding of the nature of language through comparisons of the language studied and their own.

Communities 5.1: Students use the language both within and beyond the school setting.

CORE INSTRUCTION

Warm-Up

- (5 min.) Have students study the chart shown in **Enumera los pasos y los materiales,** p. 209.

Taller del escritor

Teach

- (35 min.) Have individuals or groups of students do a chart for their own narrative, p. 209.

Wrap-Up

- (10 min.) Review the charts as a class and compare them to the chart modeled on p. 209.

OPTIONAL RESOURCES

- (15 min.) Have students translate informative reports from other textbooks, especially Science or Social Studies.
- (20 min.) Have students look for informative articles in Spanish on the Internet.

Practice Options/Homework Suggestions

- Have students read **El borrador,** p. 210.
- Have students review **El párrafo, Manual de comunicación,** R15–R16.

▲ = Advanced Learners ◆ = Slower Pace Learners ● = Special Learning Needs ■ = Heritage Speakers

Teacher's Name ______________________ Class ____________________ Date ____________

COLECCIÓN 3

El frágil medio ambiente

DAY 25 50-MINUTE LESSON PLAN

STANDARDS FOR FOREIGN LANGUAGE LEARNING: DAY 25

Taller del escritor

Communication 1.1: Students engage in conversations, provide and obtain information, express feelings and emotions, and exchange opinions.

Communication 1.2: Students understand and interpret written and spoken language on a variety of topics.

Communication 1.3: Students present information, concepts, and ideas to an audience of listeners or readers on a variety of topics.

Connections 3.1: Students reinforce and further their knowledge of other disciplines through the foreign language.

Communities 5.1: Students use the language both within and beyond the school setting.

CORE INSTRUCTION

Warm-Up

- (5 min.) Have students review **Esquema para un artículo informativo,** p. 210.

Taller del escritor

Teach

- (5 min.) Have students read **El borrador,** p. 210. See **El borrador,** *TRB,* p. 111.
- (35 min.) Have students write the first draft of **Un artículo informativo.**

Wrap-Up

- (5 min.) Have students do Additional Practice, *TRB,* p. 112.

OPTIONAL RESOURCES

- (20 min.) Have students do Extension, *TRB,* p. 112. ▲
- (10 min.) Read aloud an informational article from a Spanish-language newspaper.

Practice Options/Homework Suggestion

- Have students revise their drafts.

▲ = Advanced Learners ◆ = Slower Pace Learners ● = Special Learning Needs ■ = Heritage Speakers

El frágil medio ambiente

DAY 26 50-MINUTE LESSON PLAN

STANDARDS FOR FOREIGN LANGUAGE LEARNING: DAY 26

Taller del escritor

Communication 1.3: Students present information, concepts, and ideas to an audience of listeners or readers on a variety of topics.

Communities 5.1: Students use the language both within and beyond the school setting.

CORE INSTRUCTION

Warm-Up

- (5 min.) Have students read the questions in **Pautas de evaluación,** p. 211.

Taller del escritor

Teach

- (20 min.) Have students do **Evaluación y revisión,** Activity 1, p. 210. See **Evaluación y revisión,** *TRB,* pp. 111–112.
- (15 min.) Have students do **Evaluación y revisión,** Activity 2, p. 210.

Wrap-Up

- (10 min.) Have students compare the paragraphs in **Modelos,** pp. 211–212.

OPTIONAL RESOURCES

- (20 min.) Have students do **Publicación,** *TRB,* p. 212. ▲
- Have students compare the authors and their works in this collection, pp. 151, 160, 161, 181, and 187.

Practice Options/Homework Suggestions

- Have students use the suggestions in **Autoevaluación,** p. 210, to revise their drafts.

▲ = Advanced Learners ◆ = Slower Pace Learners ● = Special Learning Needs ■ = Heritage Speakers

El frágil medio ambiente

DAY 27 50-MINUTE LESSON PLAN

STANDARDS FOR FOREIGN LANGUAGE LEARNING: DAY 27

Taller del escritor

Communication 1.3: Students present information, concepts, and ideas to an audience of listeners or readers on a variety of topics.

Connections 3.1: Students reinforce and further their knowledge of other disciplines through the foreign language.

Communities 5.1: Students use the language both within and beyond the school setting.

CORE INSTRUCTION

Warm-Up

- (5 min.) Have students exchange their drafts with a partner and begin proofing each other's work.

Taller del escritor

Teach

- (15 min.) Have students do **Corrección de pruebas,** p. 212. See suggestions in **Corrección de pruebas,** *TRB,* p. 112.
- (15 min.) Have students do one of the activities in **Publicación,** p. 212.
- (10 min.) Have students do **Reflexión,** p. 212. See **Reflexión,** *TRB,* p. 112.

Wrap-Up

- (5 min.) Do Closure, *TRB,* p. 112.

OPTIONAL RESOURCES

- (20 min.) Do Reteaching, *TRB,* p. 112. ◆ ●

Practice Options/Homework Suggestions

- Have students do **A ver si puedo…,** pp. 213–214.

▲ = Advanced Learners ◆ = Slower Pace Learners ● = Special Learning Needs ■ = Heritage Speakers

Teacher's Name ______________________ Class ____________________ Date __________

COLECCIÓN 3

El frágil medio ambiente

DAY 28 50-MINUTE LESSON PLAN

STANDARDS FOR FOREIGN LANGUAGE LEARNING: DAY 28

A ver si puedo...

Communication 1.2: Students understand and interpret written and spoken language on a variety of topics.

Communication 1.3: Students present information, concepts, and ideas to an audience of listeners or readers on a variety of topics.

Connections 3.1: Students reinforce and further their knowledge of other disciplines through the foreign language.

Connections 3.2: Students acquire information and recognize the distinctive viewpoints that are only available through the foreign language and its cultures.

Comparisons 4.1: Students demonstrate understanding of the nature of language through comparisons of the language studied and their own.

Communities 5.1: Students use the language both within and beyond the school setting.

CORE INSTRUCTION

Warm-Up

- (5 min.) Have students review the objectives listed on the Collection Opener, p. 146.

A ver si puedo...

Review

- (10 min.) Have students do **Lectura,** Activities A and B, p. 213.
- (10 min.) Have students do **Cultura,** Activity C, p. 213.
- (10 min.) Have students do **Comunicación,** Activities D, E, F, and G, pp. 213–214.
- (10 min.) Have students do **Escritura,** Activities H, I, and J, p. 214.

Wrap–Up

- (5 min.) Answer any questions about either of the two chapter exams.

OPTIONAL RESOURCES

- (35 min.) Have students read **Enlaces literarios: La poesía del siglo XIX,** pp. 216–220. See Summary, **Antes de leer,** and Techniques for Handling the Reading, *TRB,* pp. 115–117. ▲
- (20 min.) Have students do **Comprensión del texto,** p. 221. See **Comprensión del texto,** *TRB,* p. 117. ▲
- (25 min.) Have students do **Análisis del texto,** p. 221. See **Análisis del texto,** *TRB,* pp. 117–118. ▲
- (25 min.) Have students do **Más allá del texto,** p. 221. See **Más allá del texto,** *TRB,* p. 118. ▲
- (25 min.) Have students select from activities suggested in the *TRB* for **La poesía del siglo XIX:** the Pair Work activity, Literature Link, Art Link, pp. 116–117. ▲

Practice Options/Homework Suggestions

- Have students study for the **Examen de lengua.**

▲ = Advanced Learners ◆ = Slower Pace Learners ● = Special Learning Needs ■ = Heritage Speakers

Teacher's Name ______________________ Class ______________________ Date ____________

El frágil medio ambiente

DAY 29 50-MINUTE LESSON PLAN

CORE INSTRUCTION

Assess

- (50 min.) Give **Colección 3 Examen de lengua,** *Assessment Program,* pp. 101–108.

OPTIONAL RESOURCES

- (50 min.) Give **Examen de lectura: "La fiesta del árbol," "Árbol adentro," "Meciendo," "Paisaje" y "Las abejas de bronce,"** *Assessment Program,* pp. 95–100. To allow students more time to take the exam, either **Examen** may be given over two class periods.

Practice Options/Homework Suggestions

- Have students study for the **Examen de lectura: "La fiesta del árbol," "Árbol adentro," "Meciendo," "Paisaje" y "Las abejas de bronce."**

Assessment Options

- *Assessment Program,* Performance Assessment, p. 297
- *Assessment Program,* **Midterm Exam, Examen Parcial: Lectura,** pp. 117–120
- *Assessment Program,* **Midterm Exam, Examen Parcial: Lengua,** pp. 121–123

▲ = Advanced Learners ◆ = Slower Pace Learners ● = Special Learning Needs ■ = Heritage Speakers

Teacher's Name ______________________ Class ______________________ Date ____________

El frágil medio ambiente

DAY 30 50-MINUTE LESSON PLAN

CORE INSTRUCTION

Assess

- (50 min.) Give **Examen de lectura: "La fiesta del árbol," "Árbol adentro," "Meciendo," "Paisaje" y "Las abejas de bronce,"** *Assessment Program,* pp. 95–100.

OPTIONAL RESOURCES

- (50 min.) Give **Colección 3 Examen de lengua,** *Assessment Program,* pp. 101–108. To allow students more time to take the exam, either **Examen** may be given over two class periods.

Practice Options/Homework Suggestions

- Internet (go.hrw.com, Keyword: WN3 AMBIENTE)

Assessment Options

- *Assessment Program,* Performance Assessment, p. 297
- *Assessment Program,* **Midterm Exam, Examen Parcial: Lectura,** pp. 117–120
- *Assessment Program,* **Midterm Exam, Examen Parcial: Lengua,** pp. 121–123

▲ = Advanced Learners ◆ = Slower Pace Learners ● = Special Learning Needs ■ = Heritage Speakers

Teacher's Name ______________________ Class ______________________ Date ____________

COLECCIÓN **4**

Pruebas

DAY 1 50-MINUTE LESSON PLAN

STANDARDS FOR FOREIGN LANGUAGE LEARNING: DAY 1

Lectura

Communication 1.1: Students engage in conversations, provide and obtain information, express feelings and emotions, and exchange opinions.

Communication 1.2: Students understand and interpret written and spoken language on a variety of topics.

Communication 1.3: Students present information, concepts, and ideas to an audience of listeners or readers on a variety of topics.

Connections 3.1: Students reinforce and further their knowledge of other disciplines through the foreign language.

Communities 5.1: Students use the language both within and beyond the school setting.

CORE INSTRUCTION

Warm-Up

- (5 min.) Have students read the objectives on the Collection Opener, p. 222. See Collection Overview, *TRB,* p. 122.

Lectura

Teach

- (10 min.) Have students discuss **Punto de partida,** p. 224. See **Punto de partida,** *TRB,* p. 123.
- (15 min.) In small groups, have students do **Lluvia de ideas,** p. 224. See **Lluvia de ideas,** *TRB,* p. 123.
- (15 min.) Have students begin reading aloud ***El anillo del general Macías,*** pp. 225–240. See Techniques for Handling the Reading, *TRB,* p. 124.

Wrap-Up

- (5 min.) Have students discuss **Telón de fondo,** p. 224. See **Telón de fondo,** *TRB,* p. 123.

OPTIONAL RESOURCES

- (10 min.) See Presentation Suggestions, item one, *TRB,* p. 122.
- (10 min.) See Presentation Suggestions, item two, *TRB,* p. 122.
- (10 min.) See **Vocabulario en contexto,** Group Work, *TRB,* p. 138.
- (10 min.) Read aloud ***El anillo del general Macías,*** Summary, *TRB,* p. 123. ◆ ●
- (10 min.) Have students do **Estrategias para leer,** *TRB,* p. 124. ◆ ●
- (10 min.) Play the audio recording of ***El anillo del general Macías,*** Audio CD 3, Track 9.

Practice Options/Homework Suggestions

- Internet (go.hrw.com, Keyword: WN3 PRUEBAS-LEC)
- Have students read ***El anillo del general Macías,*** pp. 225–240, including **Conoce a la escritora.**
- Have students practice the Reading Strategy, TRB, pp. 264–265. ◆ ●
- Have students study **Vocabulario esencial, *El anillo del general Macías,*** p. 295. ◆ ●

▲ = Advanced Learners ◆ = Slower Pace Learners ● = Special Learning Needs ■ = Heritage Speakers

Teacher's Name ______________________ Class ____________________ Date __________

COLECCIÓN 4

Pruebas

DAY 2 50-MINUTE LESSON PLAN

STANDARDS FOR FOREIGN LANGUAGE LEARNING: DAY 2

Lectura/Vocabulario

Communication 1.2: Students understand and interpret written and spoken language on a variety of topics.

Connections 3.1: Students reinforce and further their knowledge of other disciplines through the foreign language.

Comparisons 4.1: Students demonstrate understanding of the nature of language through comparisons of the language studied and their own.

CORE INSTRUCTION

Warm-Up

- (5 min.) Have students read the opening paragraph of **Drama** in **Elementos de literatura,** p. 250, and the definition of drama in ***Glosario de términos literarios,*** p. R8.

Lectura/Vocabulario

Teach

- (25 min.) Have students read aloud ***El anillo del general Macías,*** pp. 225–240.
- (15 min.) Have students do **Vocabulario en contexto,** Activities A and B, pp. 269–270.

Wrap-Up

- (5 min.) Have students do **Primeras impresiones,** p. 242.

OPTIONAL RESOURCES

- (10 min.) Have students do Additional Practice, *TRB,* p. 138. ◆ ●
- (30 min.) Have students do Pair Work, *TRB,* p. 138.
- (10 min.) Have pairs of students review the story in terms of **Diálogo con el texto,** p. 224. See **Diálogo con el texto,** *TRB,* p. 123.
- (15 min.) Play the audio recording of ***El anillo del general Macías,*** Audio CD 3, Track 9.

Practice Options/Homework Suggestions

- Internet (go.hrw.com, Keyword: WN3 PRUEBAS-LEC)
- Have students study **Vocabulario adicional, *El anillo del general Macías,*** *TRB,* p. 300. ◆ ●
- *Cuaderno de práctica,* Activities 1–4, pp. 67–68
- *Cuaderno de práctica,* Activities 1–2, p. 74

▲ = Advanced Learners ◆ = Slower Pace Learners ● = Special Learning Needs ■ = Heritage Speakers

Teacher's Name ______________________ Class ______________________ Date ____________

COLECCIÓN 4

Pruebas

DAY 3 50-MINUTE LESSON PLAN

STANDARDS FOR FOREIGN LANGUAGE LEARNING: DAY 3

Lectura

Communication 1.1: Students engage in conversations, provide and obtain information, express feelings and emotions, and exchange opinions.

Communication 1.2: Students understand and interpret written and spoken language on a variety of topics.

Communication 1.3: Students present information, concepts, and ideas to an audience of listeners or readers on a variety of topics.

Connections 3.2: Students acquire information and recognize the distinctive viewpoints that are only available through the foreign language and its cultures.

Communities 5.1: Students use the language both within and beyond the school setting.

CORE INSTRUCTION

Warm-Up

- (5 min.) Have students study **Así se dice** and **¿Te acuerdas?,** p. 242.

Lectura

Teach

- (15 min.) Have students do **Interpretaciones del texto,** p. 242.
- (25 min.) Have students do **Escena teatral y representación,** p. 243. See **Escena teatral y representación,** *TRB,* p. 125.

Wrap-Up

- (5 min.) Have students answer questions in **Conexiones con el texto,** p. 242.

OPTIONAL RESOURCES

- (20 min.) Have students do **Cuaderno del escritor,** p. 243. See **Cuaderno del escritor,** *TRB,* p. 124.
- (30 min.) Have students do **Arte y publicidad,** p. 243. See **Arte y publicidad,** *TRB,* p. 125. Have students discuss posters.
- (20 min.) Have students do **Para hispanohablantes,** *TRB,* p. 125. ■
- (20 min.) Have students do **Para angloparlantes,** *TRB,* p. 125.

Practice Options/Homework Suggestions

- Internet (go.hrw.com, Keyword: WN3 PRUEBAS-LEC)
- Have students do **Periodismo,** p. 243. ▲
- Have students study for **Prueba de lectura.**

▲ = Advanced Learners ◆ = Slower Pace Learners ● = Special Learning Needs ■ = Heritage Speakers

Teacher's Name ______________________ Class ____________________ Date __________

COLECCIÓN

Pruebas

DAY 4 50-MINUTE LESSON PLAN

STANDARDS FOR FOREIGN LANGUAGE LEARNING: DAY 4

Lectura/Gramática

Communication 1.2: Students understand and interpret written and spoken language on a variety of topics.

Comparisons 4.1: Students demonstrate understanding of the nature of language through comparisons of the language studied and their own.

CORE INSTRUCTION

Warm-Up

- (5 min.) Have students review ***El anillo del general Macías,*** pp. 225–240.

Lectura

Assess

- (30 min.) Give **Prueba de lectura: *El anillo del general Macías,*** Assessment Program, pp. 131–132.

Gramática

Teach

- (10 min.) Present **Las cláusulas de relativo y los pronombres relativos** (up to item three), p. 274. See **Las cláusulas de relativo y los pronombres relativos** and **Los pronombres relativos *que, quien, el que*** y ***el cual,*** *TRB,* p. 141.

Wrap-Up

- (5 min.) Have students read **¡Ojo!,** p. 274.

OPTIONAL RESOURCES

- (20 min.) Have students do Group Work, *TRB,* p. 142.

Practice Options/Homework Suggestions

- *Cuaderno de práctica,* Activities 1–4, pp. 76–78

▲ = Advanced Learners ◆ = Slower Pace Learners ● = Special Learning Needs ■ = Heritage Speakers

Pruebas

DAY 5 50-MINUTE LESSON PLAN

STANDARDS FOR FOREIGN LANGUAGE LEARNING: DAY 5

Gramática

Connections 3.1: Students reinforce and further their knowledge of other disciplines through the foreign language.

Comparisons 4.1: Students demonstrate understanding of the nature of language through comparisons of the language studied and their own.

CORE INSTRUCTION

Warm-Up

- (5 min.) Have students review **Los pronombres relativos,** p. 274.

Gramática

Teach

- (15 min.) Present **Otros pronombres relativos** and **Cláusulas especificativas y explicativas,** p. 275. See **Otros pronombres relativos** and **Cláusulas especificativas y explicativas,** *TRB,* p. 141.
- (20 min.) Have students do Activities A and B, pp. 275–276. See **Cláusulas especificativas y explicativas,** and Activities A and B, *TRB,* pp. 141–142.

Wrap-Up

- (10 min.) Have students do **Para angloparlantes,** *TRB,* p. 144. (You may suggest **"Una carta a Dios"** for this activity.)

OPTIONAL RESOURCES

- (20 min.) Have students do **Ampliación, Hoja de práctica 4-A: Más sobre el pronombre relativo *cuyo,*** *TRB,* p. 287. ◆ ●

Practice Options/Homework Suggestions

- *Cuaderno de práctica,* **Ampliación, Hoja de práctica 4-A,** p. 147 ◆ ●
- Have students read **Cultura y lengua: México,** pp. 244–248.
- Have students study for **Prueba de gramática.**

▲ = Advanced Learners ◆ = Slower Pace Learners ● = Special Learning Needs ■ = Heritage Speakers

Pruebas

DAY 6 50-MINUTE LESSON PLAN

STANDARDS FOR FOREIGN LANGUAGE LEARNING: DAY 6

Gramática/Cultura y lengua

Communication 1.2: Students understand and interpret written and spoken language on a variety of topics.

Cultures 2.1: Students demonstrate an understanding of the relationship between the practices and perspectives of the culture studied.

Cultures 2.2: Students demonstrate an understanding of the relationship between the products and perspectives of the culture studied.

Connections 3.1: Students reinforce and further their knowledge of other disciplines through the foreign language.

Connections 3.2: Students acquire information and recognize the distinctive viewpoints that are only available through the foreign language and its cultures.

CORE INSTRUCTION

Warm-Up

- (5 min.) Have students review **Las cláusulas de relativo y los pronombres relativos,** pp. 274–275.

Gramática

Assess

- (20 min.) Give **Prueba de gramática: Las cláusulas de relativo y los pronombres relativos,** *Assessment Program,* pp. 139–140.

Cultura y lengua

Teach

- (5 min.) Present **Cultura y lengua: México,** pp. 244–245. See Getting Started, *TRB,* p. 126.
- (10 min.) Have students read aloud **Cultura y lengua,** pp. 244–245.
- (5 min.) Have groups of students do History Link, *TRB,* p. 126.

Wrap-Up

- (5 min.) Have students discuss the map information, p. 244.

OPTIONAL RESOURCES

- (15 min.) See **Cultura y lengua,** Teaching Suggestions, *Video Guide,* p. 30.
- (20 min.) Have students do Thinking Critically, *TRB,* p. 126. ▲
- (20 min.) Have students do Art Link, *TRB,* p. 126.

Practice Options/Homework Suggestions

- Internet (go.hrw.com, Keyword: WN3 PRUEBAS-CYL)
- Have students read **Cultura y lengua,** pp. 246–248.

Assessment Options

- *Assessment Program,* **Prueba de gramática 4-A: Más sobre el pronombre relativo** ***cuyo,*** p. 268

▲ = Advanced Learners ◆ = Slower Pace Learners ● = Special Learning Needs ■ = Heritage Speakers

Teacher's Name ______________________ Class ____________________ Date __________

COLECCIÓN 4

Pruebas

DAY 7 50-MINUTE LESSON PLAN

STANDARDS FOR FOREIGN LANGUAGE LEARNING: DAY 7

Cultura y lengua

Communication 1.3: Students present information, concepts, and ideas to an audience of listeners or readers on a variety of topics.

Connections 3.1: Students reinforce and further their knowledge of other disciplines through the foreign language.

Connections 3.2: Students acquire information and recognize the distinctive viewpoints that are only available through the foreign language and its cultures.

Comparisons 4.2: Students demonstrate understanding of the concept of culture through comparisons of the cultures studied and their own.

Communities 5.1: Students use the language both within and beyond the school setting.

CORE INSTRUCTION

Warm-Up

- (5 min.) Have students discuss Math Link, *TRB,* p. 127.

Cultura y lengua

Teach

- (5 min.) Present **Así se dice,** p. 248. See **Así se dice,** *TRB,* p. 127.
- (15 min.) Have pairs of students do **Actividad,** p. 248.
- (10 min.) See **Cultura y lengua,** Teaching Suggestions, **Antes de ver,** *Video Guide,* p. 30. Have students select people, places, or events for presentation.
- (10 min.) Show **Cultura y lengua: México,** *Video Program* (Videocassette 2). See **Cultura y lengua,** Teaching Suggestions, **Mientras lo ves,** *Video Guide,* p. 30.

Wrap-Up

- (5 min.) Have students do item one in **Cultura y lengua,** Teaching Suggestions, **Después de ver,** *Video Guide,* p. 30.

OPTIONAL RESOURCES

- (20 min.) Have students do the first Thinking Critically, *TRB,* p. 127. ▲ ■
- (10 min.) Have students do **Hoja de actividades 2,** *Video Guide,* p. 34.
- (20 min.) Have students do item two in **Cultura y lengua,** Teaching Suggestions, **Después de ver,** *Video Guide,* p. 30. ▲
- (20 min.) Have students do item three in **Cultura y lengua,** Teaching Suggestions, **Después de ver,** *Video Guide,* p. 30.

Practice Options/Homework Suggestions

- Internet (go.hrw.com, Keyword: WN3 PRUEBAS-CYL)
- Have students study for **Prueba de cultura.**

▲ = Advanced Learners ◆ = Slower Pace Learners ● = Special Learning Needs ■ = Heritage Speakers

Pruebas

DAY 8 50-MINUTE LESSON PLAN

STANDARDS FOR FOREIGN LANGUAGE LEARNING: DAY 8

Cultura y lengua/Lectura

Communication 1.2: Students understand and interpret written and spoken language on a variety of topics.

Connections 3.1: Students reinforce and further their knowledge of other disciplines through the foreign language.

CORE INSTRUCTION

Warm-Up

- (5 min.) Have students review **Cultura y lengua,** pp. 244–248.

Cultura y lengua

Assess

- (20 min.) Give **Prueba de cultura: La Revolución mexicana,** *Assessment Program,* p. 143.

Lectura

Teach

- (15 min.) Have students do **Punto de partida,** p. 252. See **Punto de partida,** *TRB,* p. 131.
- (5 min.) Have students read **Elementos de literatura,** p. 252. See **Elementos de literatura,** *TRB,* p. 131.

Wrap-Up

- (5 min.) Have students review **Tema,** p. 252, and ***Glosario de términos literarios,*** p. R13.

OPTIONAL RESOURCES

- (5 min.) Read aloud **"Cajas de cartón,"** Summary, *TRB,* p. 131. ◆ ●

Practice Options/Homework Suggestions

- Internet (go.hrw.com, Keyword: WN3 PRUEBAS-CYL)
- Have students read **Estrategias para leer,** p. 249. ◆ ●

▲ = Advanced Learners ◆ = Slower Pace Learners ● = Special Learning Needs ■ = Heritage Speakers

Pruebas

DAY 9 50-MINUTE LESSON PLAN

STANDARDS FOR FOREIGN LANGUAGE LEARNING: DAY 9

Lectura

Communication 1.1: Students engage in conversations, provide and obtain information, express feelings and emotions, and exchange opinions.

Communication 1.2: Students understand and interpret written and spoken language on a variety of topics.

Connections 3.1: Students reinforce and further their knowledge of other disciplines through the foreign language.

Connections 3.2: Students acquire information and recognize the distinctive viewpoints that are only available through the foreign language and its cultures.

CORE INSTRUCTION

Warm-Up

- (5 min.) Have students read **Conoce al escritor,** p. 259.

Lectura

Teach

- (25 min.) Have students read aloud **"Cajas de cartón,"** pp. 253–258. See Techniques for Handling the Reading, *TRB,* p. 131.
- (15 min.) Have students read aloud **Un héroe del pueblo** and **Conoce al escritor,** p. 259.

Wrap-Up

- (5 min.) Have students review information on **Literatura y estudios sociales,** *TRB,* p. 132.

OPTIONAL RESOURCES

- (20 min.) Have students do Art Link, *TRB,* p. 131. ▲
- (30 min.) Have students practice the Reading Strategy, *TRB,* pp. 266–267. ▲ ◆ ●

Practice Options/Homework Suggestions

- Internet (go.hrw.com, Keyword: WN3 PRUEBAS-LEC)
- Have students study **Vocabulario esencial, "Cajas de cartón,"** p. 295. ◆ ●

▲ = Advanced Learners ◆ = Slower Pace Learners ● = Special Learning Needs ■ = Heritage Speakers

Teacher's Name ______________________ Class ______________________ Date ____________

Pruebas

DAY 10 50-MINUTE LESSON PLAN

STANDARDS FOR FOREIGN LANGUAGE LEARNING: DAY 10

Lectura

Communication 1.2: Students understand and interpret written and spoken language on a variety of topics.

Communication 1.3: Students present information, concepts, and ideas to an audience of listeners or readers on a variety of topics.

Connections 3.2: Students acquire information and recognize the distinctive viewpoints that are only available through the foreign language and its cultures.

Comparisons 4.2: Students demonstrate understanding of the concept of culture through comparisons of the cultures studied and their own.

CORE INSTRUCTION

Warm-Up

- (5 min.) Have students do **Repaso del texto,** p. 260.

Lectura

Teach

- (10 min.) Have students read aloud **"Cajas de cartón,"** pp. 253–258.
- (10 min.) Go over students' semantic webs from **Repaso del texto,** p. 260. Have volunteers write their webs on the chalkboard. See **Repaso del texto,** *TRB,* p. 132.
- (20 min.) Have students do **Primeras impresiones,** p. 260. See **Primeras impresiones,** *TRB,* p. 132.

Wrap-Up

- (5 min.) Have students discuss the author's attitude toward Panchito in **"Cajas de cartón."**

OPTIONAL RESOURCES

- (20 min.) Have students do **Para hispanohablantes,** *TRB,* p. 133. ■
- (20 min.) Have students do **Para angloparlantes,** *TRB,* p. 133.

Practice Options/Homework Suggestions

- Internet (go.hrw.com, Keyword: WN3 PRUEBAS-LEC)
- *Cuaderno de práctica,* Activities 1–4, pp. 69–70
- Have students study **Vocabulario adicional, "Cajas de cartón,"** *TRB,* p. 300. ◆ ●
- *Advanced Literature Preparation Program,* pp. 123–128 ▲

▲ = Advanced Learners ◆ = Slower Pace Learners ● = Special Learning Needs ■ = Heritage Speakers

Pruebas

DAY 11 50-MINUTE LESSON PLAN

STANDARDS FOR FOREIGN LANGUAGE LEARNING: DAY 11

Lectura

Communication 1.1: Students engage in conversations, provide and obtain information, express feelings and emotions, and exchange opinions.

Communication 1.2: Students understand and interpret written and spoken language on a variety of topics.

Communication 1.3: Students present information, concepts, and ideas to an audience of listeners or readers on a variety of topics.

Communities 5.1: Students use the language both within and beyond the school setting.

CORE INSTRUCTION

Warm-Up

- (5 min.) Have students study **Así se dice** and **¿Te acuerdas?,** p. 260.

Lectura

Teach

- (10 min.) Have students do **Conexiones con el texto,** p. 260.
- (25 min.) Have students do **Cuaderno del escritor,** p. 261. See **Cuaderno del escritor,** *TRB,* p. 132.

Wrap-Up

- (10 min.) Have groups volunteer their ideas and opinions from **Cuaderno del escritor.**

OPTIONAL RESOURCES

- (30 min.) Have students do **Escribe un folleto,** p. 261. See **Escribe un folleto,** *TRB,* p. 133.
- (30 min.) Have students do **La literatura y la historia actual,** p. 261. See **La literatura y la historia actual,** *TRB,* p. 132.

Practice Options/Homework Suggestions

- Internet (go.hrw.com, Keyword: WN3 PRUEBAS-LEC)

▲ = Advanced Learners ◆ = Slower Pace Learners ● = Special Learning Needs ■ = Heritage Speakers

Pruebas

DAY 12 50-MINUTE LESSON PLAN

STANDARDS FOR FOREIGN LANGUAGE LEARNING: DAY 12

Lectura/Vocabulario

Communication 1.2: Students understand and interpret written and spoken language on a variety of topics.

Comparisons 4.1: Students demonstrate understanding of the nature of language through comparisons of the language studied and their own.

CORE INSTRUCTION

Warm-Up

- (5 min.) Have students review the expressions in **Así se dice,** p. 260.

Lectura

Teach

- (10 min.) Have students do **Traducción,** p. 261. See **Traducción,** *TRB,* pp. 132–133.

Vocabulario

Teach

- (15 min.) Have students do **Vocabulario en contexto,** Activities C and D, pp. 270–271.
- (15 min.) Have students do Activity E, p. 271. You may wish to use Audio CD 2, Track 6. See script, *TRB,* pp. 139.

Wrap-Up

- (5 min.) Have students make hypothetical statements from the point of view of one of the characters in **"Cajas de cartón."** Have them use expressions from **Así se dice,** p. 260.

OPTIONAL RESOURCES

- (15 min.) Have students read **A leer por tu cuenta: "Los dos reyes y los dos laberintos,"** pp. 265–266, and **Conoce al escritor,** p. 267. See *TRB,* pp. 135–136. ▲
- (10 min.) Show the video **Lectura: "Borges y yo"** (Videocassette 2). ▲
- (30 min.) See **Lectura,** Teaching Suggestions, *Video Guide,* p. 29. ▲
- (15 min.) Have students do **Hoja de actividades 1,** *Video Guide,* p. 33. ▲
- (10 min.) Have students do **Para hispanohablantes,** *TRB,* p. 136. ■
- (10 min.) Have students do **Para angloparlantes,** *TRB,* p. 136.
- (10 min.) Have students study **Vocabulario adicional, "Los dos reyes y los dos laberintos,"** *TRB,* p. 300. ◆ ●
- Have students practice the Reading Strategy, *TRB,* p. 268. ◆ ●

Practice Options/Homework Suggestions

- Internet (go.hrw.com, Keyword: WN3 PRUEBAS-LEC)
- *Cuaderno de práctica,* Activity 3, p. 75
- *Cuaderno de práctica,* pp. 71–73
- *Cuaderno de práctica,* **Vocabulario adicional,** p. 146
- Have students study for **Prueba de lectura.**

▲ = Advanced Learners ◆ = Slower Pace Learners ● = Special Learning Needs ■ = Heritage Speakers

Pruebas

DAY 13 50-MINUTE LESSON PLAN

STANDARDS FOR FOREIGN LANGUAGE LEARNING: DAY 13

Lectura/Elementos de literatura

Communication 1.1: Students engage in conversations, provide and obtain information, express feelings and emotions, and exchange opinions.

Communication 1.2: Students understand and interpret written and spoken language on a variety of topics.

Connections 3.1: Students reinforce and further their knowledge of other disciplines through the foreign language.

Connections 3.2: Students acquire information and recognize the distinctive viewpoints that are only available through the foreign language and its cultures.

Communities 5.1: Students use the language both within and beyond the school setting.

Communities 5.2: Students show evidence of becoming life-long learners by using the language for personal enjoyment and enrichment.

CORE INSTRUCTION

Warm-Up

- (5 min.) Have students review **"Cajas de cartón,"** pp. 253–258.

Lectura

Assess

- (30 min.) Give **Prueba de lectura: "Cajas de cartón,"** *Assessment Program,* pp. 133–134.

Elementos de literatura

Teach

- (10 min.) Have students begin reading **Elementos de literatura,** pp. 250–251.

Wrap-Up

- (5 min.) Have students review the material in **Elementos de literatura,** pp. 250–251.

OPTIONAL RESOURCES

- (30 min.) Have students discuss Applying the Element, *TRB,* p. 129.
- (5 min.) Have students do History Link, *TRB,* p. 129.
- (20 min.) Have students do Community Link, *TRB,* p. 130.
- (15 min.) Have students do Art Link, *TRB,* p. 130.
- (20 min.) Have students select from Kinesthetic, Drama, Music, or Critical Thinking, *TRB,* pp. 129–130.

Practice Options/Homework Suggestions

- Internet (go.hrw.com, Keyword: WN3 PRUEBAS-LEC)
- Have students finish reading **Elementos de literatura,** pp. 250–251.

Assessment Options

- *Assessment Program,* **Prueba de lectura: "Los dos reyes y los dos laberintos,"** pp. 135–136
- *Assessment Program,* **Prueba de vocabulario adicional,** p. 267

▲ = Advanced Learners ◆ = Slower Pace Learners ● = Special Learning Needs ■ = Heritage Speakers

Pruebas

DAY 14 50-MINUTE LESSON PLAN

STANDARDS FOR FOREIGN LANGUAGE LEARNING: DAY 14

Vocabulario

Communication 1.1: Students engage in conversations, provide and obtain information, express feelings and emotions, and exchange opinions.

Communication 1.3: Students present information, concepts, and ideas to an audience of listeners or readers on a variety of topics.

Connections 3.1: Students reinforce and further their knowledge of other disciplines through the foreign language.

Comparisons 4.1: Students demonstrate understanding of the nature of language through comparisons of the language studied and their own.

Communities 5.1: Students use the language both within and beyond the school setting.

CORE INSTRUCTION

Warm-Up

- (5 min.) Have students answer the questions in Additional Practice, *TRB*, p. 130.

Vocabulario

Teach

- (5 min.) Present **Mejora tu vocabulario: Los regionalismos,** p. 271. See **Mejora tu vocabulario: Los regionalismos,** *TRB*, p. 139.
- (15 min.) Have students do Activity F, p. 272. See Activity F, *TRB*, pp. 139–140.
- (15 min.) Have pairs of students do Activity G, p. 272.

Wrap-Up

- (10 min.) Have students do Extension, *TRB*, p. 140.

OPTIONAL RESOURCES

- (20 min.) Have students do Activity I, p. 273.
- (20 min.) Have students do **Para hispanohablantes,** *TRB*, p. 140. ■
- (10 min.) Have students do **Para angloparlantes,** *TRB*, p. 140.

Practice Options/Homework Suggestions

- Have students do Activity H, p. 273.
- *Cuaderno de práctica,* Activity 4, p. 75
- Have students study **Vocabulario esencial, Mejora tu vocabulario,** p. 295 ◆ ●
- Have students study for **Prueba de vocabulario.**

▲ = Advanced Learners ◆ = Slower Pace Learners ● = Special Learning Needs ■ = Heritage Speakers

Teacher's Name ______________ Class ______________ Date ______________

COLECCIÓN **4**

Pruebas

DAY 15 50-MINUTE LESSON PLAN

STANDARDS FOR FOREIGN LANGUAGE LEARNING: DAY 15

Vocabulario/Gramática

Communication 1.2: Students understand and interpret written and spoken language on a variety of topics.

Comparisons 4.1: Students demonstrate understanding of the nature of language through comparisons of the language studied and their own.

CORE INSTRUCTION

Warm-Up

- (5 min.) Have students review **Vocabulario,** pp. 269–273.

Vocabulario

Assess

- (20 min.) Give **Prueba de vocabulario,** *Assessment Program,* pp. 137–138.

Gramática

Teach

- (20 min.) Present **Los usos de los pronombres relativos,** pp. 276–278.

Wrap-Up

- (5 min.) Review the material by asking students the questions that appear in **Los usos de los pronombres relativos,** *TRB,* p. 142.

OPTIONAL RESOURCES

- (20 min.) Have students do **Para angloparlantes,** *TRB,* p. 144.

Practice Options/Homework Suggestions

- *Cuaderno de práctica,* Activities 5–8, pp. 79–81
- *Advanced Placement Literature Preparation Book,* pp. 129–131 ▲

▲ = Advanced Learners ◆ = Slower Pace Learners ● = Special Learning Needs ■ = Heritage Speakers

Pruebas

STANDARDS FOR FOREIGN LANGUAGE LEARNING: DAY 16

Gramática

Communication 1.2: Students understand and interpret written and spoken language on a variety of topics.

Communication 1.3: Students present information, concepts, and ideas to an audience of listeners or readers on a variety of topics.

Comparisons 4.1: Students demonstrate understanding of the nature of language through comparisons of the language studied and their own.

Communities 5.1: Students use the language both within and beyond the school setting.

CORE INSTRUCTION

Warm-Up

- (5 min.) Have students review **Los usos de los pronombres relativos,** pp. 276–278.

Gramática

Teach

- (15 min.) Have students do Activity C, p. 279.
- (10 min.) Have students do Activity D, p. 279.
- (15 min.) Have students do Activity E, p. 280.

Wrap-Up

- (5 min.) Have students review **Cláusulas especificativas y explicativas,** *TRB,* p. 141.

OPTIONAL RESOURCES

- (20 min.) Have students do **Para hispanohablantes,** *TRB,* p. 144. ■

Practice Options/Homework Suggestions

- *Cuaderno de práctica,* Activities 9–11, pp. 82–83

▲ = Advanced Learners ◆ = Slower Pace Learners ● = Special Learning Needs ■ = Heritage Speakers

Pruebas

STANDARDS FOR FOREIGN LANGUAGE LEARNING: DAY 17

Gramática

Communication 1.2: Students understand and interpret written and spoken language on a variety of topics.

Communication 1.3: Students present information, concepts, and ideas to an audience of listeners or readers on a variety of topics.

Communities 5.1: Students use the language both within and beyond the school setting.

CORE INSTRUCTION

Warm-Up

- (5 min.) Have students review the chart on p. 278.

Gramática

Teach

- (10 min.) Have students do Activity F, pp. 280–281.
- (10 min.) Have students do Activity G, p. 281.
- (15 min.) Have students do Activity H, p. 281.

Wrap-Up

- (10 min.) Have student volunteers read aloud examples from Activity H, p. 281.

OPTIONAL RESOURCES

- (20 min.) Have students do Activity I, p. 281.

Practice Options/Homework Suggestions

- *Cuaderno de práctica,* Activities 12–13, p. 84
- Have students study for **Prueba de gramática.**

▲ = Advanced Learners ◆ = Slower Pace Learners ● = Special Learning Needs ■ = Heritage Speakers

Teacher's Name ______________________ Class ______________________ Date ____________

COLECCIÓN 4

Pruebas

DAY 18 50-MINUTE LESSON PLAN

STANDARDS FOR FOREIGN LANGUAGE LEARNING: DAY 18

Gramática

Connections 3.1: Students reinforce and further their knowledge of other disciplines through the foreign language.

Comparisons 4.1: Students demonstrate understanding of the nature of language through comparisons of the language studied and their own.

CORE INSTRUCTION

Warm-Up

- (5 min.) Have students review **Las cláusulas de relativo y los pronombres relativos** and **Los usos de los pronombres relativos,** pp. 274–278.

Gramática

Assess

- (30 min.) Give **Prueba de gramática: Las cláusulas de relativo y los pronombres relativos,** *Assessment Program,* pp. 139–140.

Gramática

Teach

- (10 min.) Present **Comparación y contraste,** pp. 282–283. See **Comparación y contraste,** *TRB,* p. 143.

Wrap-Up

- (5 min.) Have students do Activity A, p. 283.

OPTIONAL RESOURCES

- (10 min.) Have students compare a similar story in English and Spanish editions of a newspaper or magazine and compare the use of prepositions.

Practice Options/Homework Suggestions

- *Cuaderno de práctica,* Activities 14–15, p. 85
- Have students study for **Prueba de comparación y contraste.**

▲ = Advanced Learners ◆ = Slower Pace Learners ● = Special Learning Needs ■ = Heritage Speakers

Teacher's Name ______________ Class ______________ Date ______________

COLECCIÓN 4

Pruebas

DAY 19 50-MINUTE LESSON PLAN

STANDARDS FOR FOREIGN LANGUAGE LEARNING: DAY 19

Gramática

Connections 3.1: Students reinforce and further their knowledge of other disciplines through the foreign language.

Comparisons 4.1: Students demonstrate understanding of the nature of language through comparisons of the language studied and their own.

CORE INSTRUCTION

Warm-Up

- (5 min.) Have students review **Comparación y contraste,** pp. 282–283.

Gramática
Teach

- (10 min.) Have students do Activity B, p. 283.
- (10 min.) Have students do Activity C, p. 283.

Gramática
Assess

- (20 min.) Give **Prueba de comparación y contraste,** *Assessment Program,* pp. 141–142.

Wrap-Up

- (5 min.) Have students review differences in relative clauses between Spanish and English.

OPTIONAL RESOURCES

- (10 min.) Have students do **Las cláusulas de relativo en español e inglés,** *TRB,* p. 143.

Practice Options/Homework Suggestions

- Have students read **Panorama cultural,** pp. 262–264.

▲ = Advanced Learners ◆ = Slower Pace Learners ● = Special Learning Needs ■ = Heritage Speakers

Pruebas

STANDARDS FOR FOREIGN LANGUAGE LEARNING: DAY 20

Panorama cultural

Communication 1.1: Students engage in conversations, provide and obtain information, express feelings and emotions, and exchange opinions.

Communication 1.2: Students understand and interpret written and spoken language on a variety of topics.

Cultures 2.1: Students demonstrate an understanding of the relationship between the practices and perspectives of the culture studied.

Connections 3.2: Students acquire information and recognize the distinctive viewpoints that are only available through the foreign language and its cultures.

CORE INSTRUCTION

Warm-Up

- (5 min.) Have pairs of students discuss their answers to the topic question in **Panorama cultural,** p. 262.

Panorama cultural

Teach

- (10 min.) Have students discuss the introduction of **Panorama cultural,** p. 262. See Summary and Presentation, **Panorama cultural,** *TRB,* p. 134.
- (20 min.) Show **Panorama cultural,** *Video Program* (Videocassette 2). See **Panorama cultural,** Teaching Suggestions, **Mientras lo ves,** *Video Guide,* p. 31.
- (10 min.) Have students do Activities A–D, p. 264.

Wrap-Up

- (5 min.) Have students answer the question for item one, **Panorama cultural,** Teaching Suggestions, **Después de ver,** *Video Guide,* p. 31.

OPTIONAL RESOURCES

- (5 min.) Play Audio CD 2, Tracks 1–4, and have students listen to the interviews.
- (20 min.) Have students do **Panorama cultural,** Teaching Suggestions, **Antes de ver,** *Video Guide,* p. 31.
- (15 min.) Have students do **Hoja de actividades 3,** *Video Guide,* p. 35.
- (15 min.) Have students do Activity E, p. 264.
- (15 min.) Have students do Activity F, p. 264. You may use Audio CD 2, Track 5. See script, *TRB,* p. 134.

Practice Options/Homework Suggestions

- Internet (go.hrw.com, Keyword: WN3 PRUEBAS-CYO)
- Have students read **Comunidad y oficio,** p. 268.

▲ = Advanced Learners ◆ = Slower Pace Learners ● = Special Learning Needs ■ = Heritage Speakers

Teacher's Name ______________________ Class ______________________ Date ______________

Pruebas

DAY 21 50-MINUTE LESSON PLAN

STANDARDS FOR FOREIGN LANGUAGE LEARNING: DAY 21

Comunidad y oficio/Ortografía

Communication 1.3: Students present information, concepts, and ideas to an audience of listeners or readers on a variety of topics.

Comparisons 4.1: Students demonstrate understanding of the nature of language through comparisons of the language studied and their own.

Communities 5.1: Students use the language both within and beyond the school setting.

Communities 5.2: Students show evidence of becoming life-long learners by using the language for personal enjoyment and enrichment.

CORE INSTRUCTION

Warm-Up

- (5 min.) Have students do **Investigaciones,** Activity A, p. 268.

Comunidad y oficio

Teach

- (5 min.) Show **Comunidad y oficio,** *Video Program* (Videocassette 2). See **Comunidad y oficio,** Teaching Suggestions, **Mientras lo ves,** *Video Guide,* p. 32.
- (15 min.) Have students do **Hoja de actividades 4,** *Video Guide,* p. 36.

Ortografía

Teach

- (15 min.) Present **Las letras *m* y *n*,** p. 284. See **Las letras *m* y *n*,** *TRB,* p. 145.
- (5 min.) Have students do Activity A, p. 284.

Wrap-Up

- (5 min.) Have students do Additional Practice, *TRB,* p. 145.

OPTIONAL RESOURCES

- (30 min.) Have students do **Investigaciones,** Activity B, p. 268. See **Investigaciones,** *TRB,* p. 137.
- (5 min.) Have students do **Comunidad y oficio,** Teaching Suggestions, **Después de ver,** item two, *Video Guide,* p. 32.
- (10 min.) Have students do **Comunidad y oficio,** Teaching Suggestions, **Antes de ver,** *Video Guide,* p. 32
- (30 min.) Have students do Pair Work, *TRB,* p. 137.
- Have students do Community Link, *TRB,* p. 137. ▲ ■
- (15 min.) Have students do the first Thinking Critically activity, *TRB,* p. 137. ▲
- (15 min.) Have students do **Ampliación, Hoja de práctica 4-B, Más grupos consonánticos,** *TRB,* p. 288. ◆ ●

Practice Options/Homework Suggestions

- *Cuaderno de práctica,* Activities 1–3, p. 86
- *Cuaderno de práctica,* **Ampliación, Hoja de práctica 4-B,** 1–2, p. 148 ◆ ●
- *Advanced Placement Literature Preparation Book,* pp. 135–137 ▲

▲ = Advanced Learners ◆ = Slower Pace Learners ● = Special Learning Needs ■ = Heritage Speakers

Teacher's Name ______________________ Class ______________________ Date ____________

COLECCIÓN 4

Pruebas

DAY 22 50-MINUTE LESSON PLAN

STANDARDS FOR FOREIGN LANGUAGE LEARNING: DAY 22

Ortografía

Communication 1.1: Students engage in conversations, provide and obtain information, express feelings and emotions, and exchange opinions.

Communication 1.2: Students understand and interpret written and spoken language on a variety of topics.

Comparisons 4.1: Students demonstrate understanding of the nature of language through comparisons of the language studied and their own.

Communities 5.1: Students use the language both within and beyond the school setting.

CORE INSTRUCTION

Warm-Up

- (5 min.) Have students do Activity B, p. 285.

Ortografía

Teach

- (10 min.) Have students do Activity C, p. 285. See second **¡Ojo!,** p. 284, and **Las letras *m* y *n*,** *TRB,* p. 145.
- (15 min.) Present **La acentuación,** pp. 285–286. See **La acentuación,** *TRB,* pp. 145–146.
- (15 min.) Have students do Activities D and E, p. 287.

Wrap-Up

- (5 min.) Present **¡Ojo!,** p. 286, and Language Note, *TRB,* p. 146.

OPTIONAL RESOURCES

- (10 min.) Have students do Activity F, p. 287.
- (10 min.) Have students do Group Work, *TRB,* p. 146. ▲
- (20 min.) Have students do **Para hispanohablantes,** *TRB,* p. 147. ■
- (20 min.) Have students do **Para angloparlantes,** *TRB,* p. 147.

Practice Options/Homework Suggestions

- *Cuaderno de práctica,* Activities 4–6, p. 87
- Have students study for **Prueba de ortografía.**

Assessment Options

- *Assessment Program,* **Prueba de ortografía 4-B: Más grupos consonánticos,** p. 269

▲ = Advanced Learners ◆ = Slower Pace Learners ● = Special Learning Needs ■ = Heritage Speakers

Pruebas

DAY 23 50-MINUTE LESSON PLAN

STANDARDS FOR FOREIGN LANGUAGE LEARNING: DAY 23

Ortografía/Taller del escritor

Communication 1.2: Students understand and interpret written and spoken language on a variety of topics.

Connections 3.1: Students reinforce and further their knowledge of other disciplines through the foreign language.

Communities 5.1: Students use the language both within and beyond the school setting.

CORE INSTRUCTION

Warm-Up

- (5 min.) Have students review **Ortografía,** pp. 284–287.

Ortografía
Teach

- (10 min.) Give **Dictado,** Activities A and B, p. 287. You may wish to use Audio CD 2, Tracks 7–8. See scripts, *TRB,* p. 147.

Ortografía
Assess

- (15 min.) Give **Prueba de ortografía,** *Assessment Program,* p. 144.

Taller del escritor
Teach

- (5 min.) Introduce **Taller del escritor,** p. 288. Have students read **Ensayo sobre problemas y soluciones,** p. 288.
- (10 min.) Have students do **Antes de escribir: Cuaderno del escritor,** p. 288. See **Antes de escribir,** *TRB,* p. 148.

Wrap-Up

- (5 min.) Have students go over **Objetivos de un ensayo sobre problemas y soluciones,** p. 288.

OPTIONAL RESOURCES

- (20 min.) Have students read and discuss **Examina los medios de comunicación,** p. 288.

Practice Options/Homework Suggestions

- Have students choose the theme of their essay.
- Have students read **Explora un problema y su solución,** p. 289.

▲ = Advanced Learners ◆ = Slower Pace Learners ● = Special Learning Needs ■ = Heritage Speakers

Pruebas

DAY 24 50-MINUTE LESSON PLAN

STANDARDS FOR FOREIGN LANGUAGE LEARNING: DAY 24

Taller del escritor

Communication 1.2: Students understand and interpret written and spoken language on a variety of topics.

Communication 1.3: Students present information, concepts, and ideas to an audience of listeners or readers on a variety of topics.

Communities 5.1: Students use the language both within and beyond the school setting.

CORE INSTRUCTION

Warm-Up

- (5 min.) Have students study the **Problema** chart, p. 289.

Taller del escritor

Teach

- (20 min.) Have groups of students or individuals do a chart for their own essay, p. 289.
- (20 min.) Have students do **Preguntas para encontrar soluciones,** p. 289.

Wrap-Up

- (5 min.) Have students review the charts and compare them to the chart modeled on p. 289.

OPTIONAL RESOURCES

- (15 min.) Have students read editorial sections of Spanish-language newspapers for essays about local or national issues. See **Ensayo sobre problemas y soluciones,** *TRB,* p. 148.

Practice Options/Homework Suggestions

- Have students read **Busca y respalda la mejor solución,** p. 290.

▲ = Advanced Learners ◆ = Slower Pace Learners ● = Special Learning Needs ■ = Heritage Speakers

Teacher's Name ______________________ Class ______________________ Date ______________

COLECCIÓN

Pruebas

DAY 25 50-MINUTE LESSON PLAN

STANDARDS FOR FOREIGN LANGUAGE LEARNING: DAY 25

Taller del escritor

Communication 1.2: Students understand and interpret written and spoken language on a variety of topics.

Communication 1.3: Students present information, concepts, and ideas to an audience of listeners or readers on a variety of topics.

Connections 3.1: Students reinforce and further their knowledge of other disciplines through the foreign language.

CORE INSTRUCTION

Warm-Up

- (5 min.) Have students read **Hecho contra opinión,** p. 290.

Taller del escritor

Teach

- (10 min.) Have students read **El borrador,** p. 290. See **El borrador,** *TRB,* p. 148.
- (30 min.) Have students write the first draft of **Un ensayo sobre problemas y soluciones.**

Wrap-Up

- (5 min.) Have students read the chart titled **Esquema para un ensayo sobre problemas y soluciones,** p. 290.

OPTIONAL RESOURCES

- (15 min.) Have students read and compare a news article with an editorial.
- Have students compare the authors in this collection and their works, pp. 240, 259, and 267. ▲

Practice Options/Homework Suggestion

- Have students revise their drafts.

▲ = Advanced Learners ◆ = Slower Pace Learners ● = Special Learning Needs ■ = Heritage Speakers

Teacher's Name ______________________ Class ______________________ Date ______________

Pruebas

DAY 26 50-MINUTE LESSON PLAN

STANDARDS FOR FOREIGN LANGUAGE LEARNING: DAY 26

Taller del escritor

Communication 1.1: Students engage in conversations, provide and obtain information, express feelings and emotions, and exchange opinions.

Communication 1.2: Students understand and interpret written and spoken language on a variety of topics.

Communication 1.3: Students present information, concepts, and ideas to an audience of listeners or readers on a variety of topics.

Communities 5.1: Students use the language both within and beyond the school setting.

CORE INSTRUCTION

Warm-Up

- (5 min.) Have students do **Evaluación del ensayo sobre problemas y soluciones,** p. 291.

Taller del escritor

Teach

- (30 min.) Have students do **Evaluación y revisión,** Activity 1, p. 290.
- (10 min.) Have students compare the **Modelos,** pp. 291–292. See **Evaluación y revisión,** *TRB,* pp. 148–149.

Wrap-Up

- (5 min.) Have student volunteers share their examples of evaluations from Activity 1, pp. 290–291.

OPTIONAL RESOURCES

- (20 min.) Have students do **Publicación,** *TRB,* p. 149. ▲
- (20 min.) Have students do Reteaching, *TRB,* p. 149. ◆ ●

Practice Options/Homework Suggestions

- Have students use the suggestions in **Autoevaluación,** p. 291, to revise their drafts.

▲ = Advanced Learners ◆ = Slower Pace Learners ● = Special Learning Needs ■ = Heritage Speakers

Teacher's Name ______________________ Class ______________________ Date ____________

COLECCIÓN 4

Pruebas

DAY 27 50-MINUTE LESSON PLAN

STANDARDS FOR FOREIGN LANGUAGE LEARNING: DAY 27

Taller del escritor

Communication 1.3: Students present information, concepts, and ideas to an audience of listeners or readers on a variety of topics.

Communities 5.1: Students use the language both within and beyond the school setting.

CORE INSTRUCTION

Warm-Up

- (5 min.) Have students exchange their drafts and proofread for spelling.

Taller del escritor

Teach

- (20 min.) Have students do **Corrección de pruebas,** p. 292. See suggestions in **Corrección de pruebas,** *TRB,* p. 149.
- (15 min.) Have students do one of the activities in **Publicación,** p. 292.
- (5 min.) Have students do **Reflexión,** p. 292. Have them use expressions from **Así se dice,** p. 292. See **Reflexión,** *TRB,* p. 149.

Wrap-Up

- (5 min.) Have students do Closure, *TRB,* p. 149.

OPTIONAL RESOURCES

- (20 min.) Have students follow the Assessment criteria, *TRB,* p. 149.

Practice Options/Homework Suggestions

- Have students do **A ver si puedo...,** pp. 293–294.

▲ = Advanced Learners ◆ = Slower Pace Learners ● = Special Learning Needs ■ = Heritage Speakers

Teacher's Name ______________________ Class ______________________ Date ______________

Pruebas

DAY 28 50-MINUTE LESSON PLAN

STANDARDS FOR FOREIGN LANGUAGE LEARNING: DAY 28

A ver si puedo...

Communication 1.2: Students understand and interpret written and spoken language on a variety of topics.

Communication 1.3: Students present information, concepts, and ideas to an audience of listeners or readers on a variety of topics.

Cultures 2.1: Students demonstrate an understanding of the relationship between the practices and perspectives of the culture studied.

Connections 3.1: Students reinforce and further their knowledge of other disciplines through the foreign language.

Connections 3.2: Students acquire information and recognize the distinctive viewpoints that are only available through the foreign language and its cultures.

Comparisons 4.2: Students demonstrate understanding of the concept of culture through comparisons of the cultures studied and their own.

Communities 5.2: Students show evidence of becoming life-long learners by using the language for personal enjoyment and enrichment.

CORE INSTRUCTION

Warm-Up

- (5 min.) Have students review the objectives listed on the Collection Opener, p. 222.

A ver si puedo...

Review

- (10 min.) Have students do **Lectura,** Activities A and B, p. 293.
- (10 min.) Have students do **Cultura,** Activity C, p. 293.
- (10 min.) Have students do **Comunicación,** Activities D, E, F, and G, pp. 293–294.
- (10 min.) Have students do **Escritura,** Activities H, I, and J, p. 294.

Wrap-Up

- (5 min.) Answer any questions about either of the two chapter exams.

OPTIONAL RESOURCES

- (35 min.) Have students read **Enlaces literarios: La poesía latinoamericana del siglo XX,** pp. 296–300. See **Antes de leer,** *TRB,* pp. 152–153. ▲
- (20 min.) Have students do **Comprensión del texto,** p. 301. See **Comprensión del texto,** *TRB,* pp. 154–155. ▲
- (25 min.) Have students do **Análisis del texto,** p. 301. See **Análisis del texto,** *TRB,* p. 155. ▲
- (25 min.) Have students do **Más allá del texto,** p. 301. See **Más allá del texto,** *TRB,* p. 155. ▲
- (25 min.) Have students select from activities suggested in the *TRB,* pp. 152–154. ▲

Practice Options/Homework Suggestions

- Have students study for the **Examen de lengua.**

▲ = Advanced Learners ◆ = Slower Pace Learners ● = Special Learning Needs ■ = Heritage Speakers

Pruebas

DAY 29 50-MINUTE LESSON PLAN

CORE INSTRUCTION

Assess

- (50 min.) Give **Colección 4 Examen de lengua,** *Assessment Program,* pp. 151–155.

OPTIONAL RESOURCES

- (50 min.) Give **Examen de lectura: *El anillo del general Macías* y "Cajas de cartón,"** *Assessment Program,* pp. 145–149. To allow students more time to take the exam, either **Examen** may be given over two class periods.

Practice Options/Homework Suggestions

- Have students study for the **Examen de lectura: *El anillo del general Macías* y "Cajas de cartón."**

Assessment Options

- *Assessment Program,* Performance Assessment, p. 298

▲ = Advanced Learners ◆ = Slower Pace Learners ● = Special Learning Needs ■ = Heritage Speakers

Pruebas

DAY 30 50-MINUTE LESSON PLAN

CORE INSTRUCTION

Assess

- (50 min.) Give **Examen de lectura: *El anillo del general Macías*** y **"Cajas de cartón,"** *Assessment Program,* pp. 145–149.

OPTIONAL RESOURCES

- (50 min.) Give **Colección 4 Examen de lengua,** *Assessment Program,* pp. 151–155. To allow students more time to take the exam, either **Examen** may be given over two class periods.

Practice Options/Homework Suggestions

- Internet (go.hrw.com, Keyword: WN3 PRUEBAS)

Assessment Options

- *Assessment Program,* Performance Assessment, p. 298

▲ = Advanced Learners ◆ = Slower Pace Learners ● = Special Learning Needs ■ = Heritage Speakers

Teacher's Name ______________________ Class ______________________ Date ____________

COLECCIÓN

Mitos

DAY 1 50-MINUTE LESSON PLAN

STANDARDS FOR FOREIGN LANGUAGE LEARNING: DAY 1

Lectura

Communication 1.2: Students understand and interpret written and spoken language on a variety of topics.

Communication 1.3: Students present information, concepts, and ideas to an audience of listeners or readers on a variety of topics.

Cultures 2.1: Students demonstrate an understanding of the relationship between the practices and perspectives of the culture studied.

Cultures 2.2: Students demonstrate an understanding of the relationship between the products and perspectives of the culture studied.

Connections 3.1: Students reinforce and further their knowledge of other disciplines through the foreign language.

Comparisons 4.2: Students demonstrate understanding of the concept of culture through comparisons of the cultures studied and their own.

Communities 5.2: Students show evidence of becoming lifelong learners by using the language for personal enjoyment and enrichment.

CORE INSTRUCTION

Warm-Up

- (5 min.) Have students read the objectives on the Collection Opener, p. 302. See Collection Overview, *TRB,* p. 160.

Lectura

Teach

- (10 min.) Have students discuss **Punto de partida,** p. 304. See **Punto de partida,** *TRB,* p. 161.
- (20 min.) Have students do **Escritura libre,** p. 304.
- (10 min.) Present **Telón de fondo,** pp. 304–305. See **Telón de fondo,** *TRB,* p. 161.

Wrap-Up

- (5 min.) Ask students what myths they know and what these myths are about or what they explain.

OPTIONAL RESOURCES

- (10 min.) See Presentation Suggestions, item one, *TRB,* p. 160.
- (10 min.) See Presentation Suggestions, item two, *TRB,* p. 160.
- (10 min.) See **Vocabulario en contexto,** Group Work, *TRB,* p. 176.
- (10 min.) Read aloud **del *Popul Vuh,*** Summary, *TRB,* p. 161. ◆ ●

Practice Options/Homework Suggestions

- Internet (go.hrw.com, Keyword: WN3 MITOS-LEC)
- Have students read **del *Popul Vuh,*** pp. 306–308.
- Have students practice the Reading Strategy, TRB, p. 269. ◆ ●
- Have students study **Vocabulario esencial, del *Popul Vuh,*** p. 363. ◆ ●

▲ = Advanced Learners ◆ = Slower Pace Learners ● = Special Learning Needs ■ = Heritage Speakers

Teacher's Name ______________________ Class ______________________ Date ______________

COLECCIÓN

Mitos

DAY 2 50-MINUTE LESSON PLAN

STANDARDS FOR FOREIGN LANGUAGE LEARNING: DAY 2

Lectura/Vocabulario

Communication 1.1: Students engage in conversations, provide and obtain information, express feelings and emotions, and exchange opinions.

Communication 1.2: Students understand and interpret written and spoken language on a variety of topics.

Cultures 2.1: Students demonstrate an understanding of the relationship between the practices and perspectives of the culture studied.

Cultures 2.2: Students demonstrate an understanding of the relationship between the products and perspectives of the culture studied.

Connections 3.1: Students reinforce and further their knowledge of other disciplines through the foreign language.

CORE INSTRUCTION

Warm-Up

- (5 min.) Have students read **Elementos de literatura,** p. 305, and ***Glosario de términos literarios,*** p. R10.

Lectura

Teach

- (5 min.) Have students discuss **Elementos de literatura,** *TRB,* p. 161.
- (20 min.) Have students read aloud **del *Popul Vuh,*** pp. 306–308, and **Literatura y antropología,** p. 308. See **Literatura y antropología,** *TRB,* p. 162.

Vocabulario

Teach

- (5 min.) Have students do **Vocabulario en contexto,** Activity B, pp. 337–338. See Activity B, *TRB,* p. 176.
- (10 min.) Have students answer questions in **Interpretaciones del texto,** p. 310. See **Interpretaciones del texto,** *TRB,* p. 162.

Wrap-Up

- (5 min.) Have students discuss answers to **Primeras impresiones,** p. 310.

OPTIONAL RESOURCES

- (20 min.) See Techniques for Handling the Reading, *TRB,* p. 162. ◆ ●
- (10 min.) Have pairs of students review the story in terms of **Diálogo con el texto,** p. 305. See **Diálogo con el texto,** *TRB,* p. 162.
- (5 min.) Have students do Challenge, *TRB,* p. 176. ▲

Practice Options/Homework Suggestions

- Internet (go.hrw.com, Keyword: WN3 MITOS-LEC)
- Have students study **Vocabulario adicional, del *Popul Vuh,*** *TRB,* p. 301. ◆ ●
- *Cuaderno de práctica,* Activities 1–3, pp. 89–90
- *Advanced Placement Literature Preparation Book,* pp. 92–98 ▲

▲ = Advanced Learners ◆ = Slower Pace Learners ● = Special Learning Needs ■ = Heritage Speakers

Teacher's Name ______________ Class ______________ Date ______________

COLECCIÓN

Mitos

DAY 3 50-MINUTE LESSON PLAN

STANDARDS FOR FOREIGN LANGUAGE LEARNING: DAY 3

Lectura

Communication 1.1: Students engage in conversations, provide and obtain information, express feelings and emotions, and exchange opinions.

Communication 1.2: Students understand and interpret written and spoken language on a variety of topics.

Communication 1.3: Students present information, concepts, and ideas to an audience of listeners or readers on a variety of topics.

Comparisons 4.2: Students demonstrate understanding of the concept of culture through comparisons of the cultures studied and their own.

Communities 5.1: Students use the language both within and beyond the school setting.

CORE INSTRUCTION

Warm-Up

- (5 min.) Have students do **Hoja de actividades 1, Antes de ver,** *Video Guide,* p. 42.

Lectura

Teach

- (10 min.) Show **Popul Vuh: Leyenda de los mayas quiché,** *Video Program* (Videocassette 2). See Teaching Suggestions, **Mientras lo ves,** *Video Guide,* p. 38.
- (5 min.) Have students do **Hoja de actividades 1, Mientras lo ves,** *Video Guide,* p. 42.
- (25 min.) Have students do **Cuaderno del escritor**, p. 311, using expressions from **Así se dice,** p. 310. See **Cuaderno del escritor,** *TRB,* p. 163.

Wrap-Up

- (5 min.) Have students answer questions in **Conexiones con el texto,** p. 310.

OPTIONAL RESOURCES

- (30 min.) Have students do Teaching Suggestions, **Después de ver,** *Video Guide,* p. 38. ▲
- (20 min.) Have students do **Hoja de actividades 1, Después de ver,** *Video Guide,* p. 42.
- (25 min.) Have students do **Arte,** p. 311. See **Arte,** *TRB,* p. 163. ▲
- (20 min.) Have students do **Presentación,** p. 311. See **Presentación,** *TRB,* p. 163. ▲
- (20 min.) Have students do **Para hispanohablantes,** *TRB,* p. 163. ■
- (20 min.) Have students do **Para angloparlantes,** *TRB,* p. 163.
- (20 min.) Have students do **Repaso del texto,** p. 310. See **Repaso del texto,** *TRB,* p. 162.

Practice Options/Homework Suggestions

- Internet (go.hrw.com, Keyword: WN3 MITOS-LEC)
- Have students study for **Prueba de lectura.**

▲ = Advanced Learners ◆ = Slower Pace Learners ● = Special Learning Needs ■ = Heritage Speakers

Teacher's Name ______________________ Class ____________________ Date __________

COLECCIÓN 5

Mitos

DAY 4 50-MINUTE LESSON PLAN

STANDARDS FOR FOREIGN LANGUAGE LEARNING: DAY 4

Lectura/Gramática

Communication 1.2: Students understand and interpret written and spoken language on a variety of topics.

Connections 3.1: Students reinforce and further their knowledge of other disciplines through the foreign language.

Comparisons 4.1: Students demonstrate understanding of the nature of language through comparisons of the language studied and their own.

CORE INSTRUCTION

Warm-Up

- (5 min.) Have students review **del *Popul Vuh,*** pp. 306–308.

Lectura

Assess

- (30 min.) Give **Prueba de lectura: del *Popul Vuh,*** *Assessment Program,* pp. 165–166.

Gramática

Teach

- (5 min.) Present **Repaso de las cláusulas de relativo,** p. 342.

Wrap-Up

- (10 min.) Have students do Activity A, p. 342.

OPTIONAL RESOURCES

- (15 min.) Have students do **Repaso de las cláusulas de relativo,** *TRB,* p. 179.
- (20 min.) Have students do Cooperative Learning, *TRB,* p. 180. ◆ ●

Practice Options/Homework Suggestions

- *Cuaderno de práctica,* Activities 1–3, pp. 98–99

▲ = Advanced Learners ◆ = Slower Pace Learners ● = Special Learning Needs ■ = Heritage Speakers

Teacher's Name ______________________ Class ______________________ Date __________

COLECCIÓN 5

Mitos

DAY 5 50-MINUTE LESSON PLAN

STANDARDS FOR FOREIGN LANGUAGE LEARNING: DAY 5

Gramática

Communication 1.2: Students understand and interpret written and spoken language on a variety of topics.

Communication 1.3: Students present information, concepts, and ideas to an audience of listeners or readers on a variety of topics.

Connections 3.1: Students reinforce and further their knowledge of other disciplines through the foreign language.

Comparisons 4.1: Students demonstrate understanding of the nature of language through comparisons of the language studied and their own.

CORE INSTRUCTION

Warm-Up

- (5 min.) Have students read items one and two of **El modo en las cláusulas de relativo,** p. 343.

Gramática

Teach

- (15 min.) Present **El modo en las cláusulas de relativo,** p. 343. See **El modo en las cláusulas de relativo,** *TRB,* p. 179.
- (15 min.) Have students do Activity B, p. 344. See Activity B, *TRB,* pp. 179–180.
- (10 min.) Have students do Activity C, pp. 344–345. See Activity C, *TRB,* p. 180.

Wrap-Up

- (5 min.) Have students review **El modo en las cláusulas de relativo** and **¡Ojo!,** p. 343.

OPTIONAL RESOURCES

- (20 min.) Have students do Group Work, *TRB,* p. 180. ▲

Practice Options/Homework Suggestions

- *Cuaderno de práctica,* Activities 4–6, pp. 99–100
- *Advanced Placement Literature Preparation Book,* pp. 99–103 ▲

▲ = Advanced Learners ◆ = Slower Pace Learners ● = Special Learning Needs ■ = Heritage Speakers

Teacher's Name ______________________ Class ____________________ Date __________

COLECCIÓN 5

Mitos

DAY 6 50-MINUTE LESSON PLAN

STANDARDS FOR FOREIGN LANGUAGE LEARNING: DAY 6

Gramática

Communication 1.1: Students engage in conversations, provide and obtain information, express feelings and emotions, and exchange opinions.

Communication 1.2: Students understand and interpret written and spoken language on a variety of topics.

Communication 1.3: Students present information, concepts, and ideas to an audience of listeners or readers on a variety of topics.

Comparisons 4.1: Students demonstrate understanding of the nature of language through comparisons of the language studied and their own.

CORE INSTRUCTION

Warm-Up

- (5 min.) Have students review **¡Ojo!,** p. 343.

Gramática

Teach

- (10 min.) Have students do Activity D, p. 345.
- (10 min.) Have students do Activity E, p. 346.
- (20 min.) Have students do Activity F, p. 346.

Wrap-Up

- (5 min.) Have student volunteers read their paragraphs from Activity F, p. 346.

OPTIONAL RESOURCES

- (15 min.) Have students do Pair Work, p. 180. ◆ ●

Practice Options/Homework Suggestions

- *Cuaderno de práctica,* Activities 7–9, pp. 101–102
- Have students study for **Prueba de gramática.**

▲ = Advanced Learners ◆ = Slower Pace Learners ● = Special Learning Needs ■ = Heritage Speakers

Teacher's Name ______________________ Class ______________________ Date ____________

Mitos

DAY 7 50-MINUTE LESSON PLAN

STANDARDS FOR FOREIGN LANGUAGE LEARNING: DAY 7

Gramática/Cultura y lengua

Communication 1.3: Students present information, concepts, and ideas to an audience of listeners or readers on a variety of topics.

Connections 3.1: Students reinforce and further their knowledge of other disciplines through the foreign language.

Comparisons 4.1: Students demonstrate understanding of the nature of language through comparisons of the language studied and their own.

CORE INSTRUCTION

Warm-Up

- (5 min.) Have students review **El modo en las cláusulas de relativo,** pp. 343–346.

Gramática

Assess

- (25 min.) Give **Prueba de gramática: El modo en las cláusulas de relativo,** *Assessment Program,* pp. 173–174.

Cultura y lengua

Teach

- (15 min.) Present **Lengua: El idioma maya hoy y ayer,** p. 316.

Wrap-Up

- (5 min.) Review with students the difference between pictographs and a phonetic alphabet.

OPTIONAL RESOURCES

- (20 min.) Have students do Language Link, *TRB,* p. 165.

Practice Options/Homework Suggestions

- Internet (go.hrw.com, Keyword: WN3 MITOS-CYL)
- Have students read **Cultura y lengua: Los mayas,** pp. 314–315.

▲ = Advanced Learners ◆ = Slower Pace Learners ● = Special Learning Needs ■ = Heritage Speakers

Teacher's Name ______________________ Class ______________________ Date ____________

Mitos

DAY 8 50-MINUTE LESSON PLAN

STANDARDS FOR FOREIGN LANGUAGE LEARNING: DAY 8

Cultura y lengua

Cultures 2.1: Students demonstrate an understanding of the relationship between the practices and perspectives of the culture studied.

Cultures 2.2: Students demonstrate an understanding of the relationship between the products and perspectives of the culture studied.

Connections 3.1: Students reinforce and further their knowledge of other disciplines through the foreign language.

Connections 3.2: Students acquire information and recognize the distinctive viewpoints that are only available through the foreign language and its cultures.

CORE INSTRUCTION

Warm-Up

- (5 min.) Have students study expressions in **Así se dice,** p. 316.

Cultura y lengua

Teach

- (15 min.) Have students read aloud **Cultura y lengua: Los mayas,** pp. 312–313.
- (10 min.) Show **Cultura y lengua: Los mayas,** *Video Program* (Videocassette 2). See **Cultura y lengua,** Teaching Suggestions, **Antes de ver** and **Mientras lo ves,** *Video Guide,* p. 39.
- (15 min.) Have students do **Hoja de actividades 2,** *Video Guide,* p. 43.

Wrap-Up

- (5 min.) Have students make conjectures using the expressions in **Así se dice,** p. 316. See **Así se dice,** *TRB,* p. 165.

OPTIONAL RESOURCES

- (10 min.) Choose from among several activities in *TRB,* pp. 164–165.

Practice Options/Homework Suggestions

- Internet (go.hrw.com, Keyword: WN3 MITOS-CYL)
- Have students study for **Prueba de cultura.**
- Have students read **Estrategias para leer,** p. 332. ◆ ●
- *Advanced Placement Literature Preparation Book,* pp. 104–105 ▲

▲ = Advanced Learners ◆ = Slower Pace Learners ● = Special Learning Needs ■ = Heritage Speakers

Mitos

DAY 9 50-MINUTE LESSON PLAN

STANDARDS FOR FOREIGN LANGUAGE LEARNING: DAY 9

Cultura y lengua/Lectura

Communication 1.2: Students understand and interpret written and spoken language on a variety of topics.

Communication 1.3: Students present information, concepts, and ideas to an audience of listeners or readers on a variety of topics.

Connections 3.1: Students reinforce and further their knowledge of other disciplines through the foreign language.

Communities 5.1: Students use the language both within and beyond the school setting.

CORE INSTRUCTION

Warm-Up

- (5 min.) Have students review **Cultura y lengua,** pp. 312–316.

Cultura y lengua

Assess

- (20 min.) Give **Prueba de cultura: El reino maya,** *Assessment Program,* p. 177.

Lectura

Teach

- (20 min.) Have students read **Elementos de literatura,** p. 317. Do Applying the Element, *TRB,* p. 166.

Wrap-Up

- (5 min.) Have students do Additional Practice, *TRB,* p. 167.

OPTIONAL RESOURCES

- (30 min.) Have students do Community Link, *TRB,* pp. 166–167. ▲
- (20 min.) Have students do Group Work, *TRB,* p. 167. ◆ ●
- (20 min.) Have students do Literature Link, *TRB,* p. 167.

Practice Options/Homework Suggestions

- Internet (go.hrw.com, Keyword: WN3 MITOS-LEC)
- Have students read **Tres mitos latinoamericanos: "La historia de Quetzalcóatl"** and **"El casamiento del Sol,"** pp. 320–325.

▲ = Advanced Learners ◆ = Slower Pace Learners ● = Special Learning Needs ■ = Heritage Speakers

Teacher's Name ______________________ Class ______________________ Date ______________

COLECCIÓN 5

Mitos

DAY 10 50-MINUTE LESSON PLAN

STANDARDS FOR FOREIGN LANGUAGE LEARNING: DAY 10

Lectura

Communication 1.2: Students understand and interpret written and spoken language on a variety of topics.

Communication 1.3: Students present information, concepts, and ideas to an audience of listeners or readers on a variety of topics.

Cultures 2.1: Students demonstrate an understanding of the relationship between the practices and perspectives of the culture studied.

Cultures 2.2: Students demonstrate an understanding of the relationship between the products and perspectives of the culture studied.

Connections 3.1: Students reinforce and further their knowledge of other disciplines through the foreign language.

Connections 3.2: Students acquire information and recognize the distinctive viewpoints that are only available through the foreign language and its cultures.

CORE INSTRUCTION

Warm-Up

- (5 min.) Have students read **Punto de partida** and **Telón de fondo,** p. 318. See **Punto de partida** and **Telón de fondo,** *TRB,* pp. 168–169.

Lectura

Teach

- (20 min.) Have students read aloud **"Los primeros incas,"** pp. 326–327. See Techniques for Handling the Reading, *TRB,* p. 169.
- (15 min.) Present both **Literatura y antropología,** pp. 322 and 325. See **Literatura y antropología,** *TRB,* p. 169.

Wrap-Up

- (10 min.) Discuss with students the map, photos, and drawings, pp. 319–327.

OPTIONAL RESOURCES

- (5 min.) Read aloud **"Tres mitos latinoamericanos,"** Summary, *TRB,* p. 168. ◆ ●
- (30 min.) Have students practice the Reading Strategy, *TRB,* pp. 270–271. ▲ ◆ ●

Practice Options/Homework Suggestions

- Internet (go.hrw.com, Keyword: WN3 MITOS-LEC)
- Have students do the first activity in **Primeras impresiones,** p. 328.
- Have students study **Vocabulario esencial, "Tres mitos latinoamericanos,"** p. 363. ◆ ●
- *Cuaderno de práctica,* Activities 1–3, pp. 91–92

▲ = Advanced Learners ◆ = Slower Pace Learners ● = Special Learning Needs ■ = Heritage Speakers

Mitos

DAY 11 50-MINUTE LESSON PLAN

STANDARDS FOR FOREIGN LANGUAGE LEARNING: DAY 11

Lectura

Communication 1.2: Students understand and interpret written and spoken language on a variety of topics.

Communication 1.3: Students present information, concepts, and ideas to an audience of listeners or readers on a variety of topics.

Connections 3.1: Students reinforce and further their knowledge of other disciplines through the foreign language.

Communities 5.1: Students use the language both within and beyond the school setting.

CORE INSTRUCTION

Warm-Up

- (5 min.) In pairs, have students do **Primeras impresiones,** p. 328.

Lectura

Teach

- (20 min.) Have students do **Interpretaciones del texto,** p. 328.
- (10 min.) Have students do **Conexiones con el texto,** p. 328.
- (10 min.) Have students do **Más allá del texto,** p. 328, using the expressions in **Así se dice.** See **Más allá del texto,** *TRB,* p. 169.

Wrap-Up

- (5 min.) Review with students **¿Te acuerdas?** and the expressions in **Así se dice,** p. 328.

OPTIONAL RESOURCES

- (30 min.) Have students do **Drama,** p. 329. See **Drama,** *TRB,* p. 170.
- (30 min.) Have students do **Carteles,** p. 329. See **Carteles,** *TRB,* p. 170.
- Have students do **Escritura Creativa,** p. 329.

Practice Options/Homework Suggestions

- Internet (go.hrw.com, Keyword: WN3 MITOS-LEC)
- Have students study **Vocabulario adicional, "Tres mitos latinoamericanos,"** *TRB,* p. 301. ◆ ●
- Have students study for **Prueba de lectura.**
- *Advanced Placement Literature Preparation Book,* pp. 106–108 ▲

▲ = Advanced Learners ◆ = Slower Pace Learners ● = Special Learning Needs ■ = Heritage Speakers

Teacher's Name ______________________ Class ______________________ Date ______________

COLECCIÓN

Mitos

DAY 12 50-MINUTE LESSON PLAN

STANDARDS FOR FOREIGN LANGUAGE LEARNING: DAY 12

Lectura/Vocabulario

Communication 1.2: Students understand and interpret written and spoken language on a variety of topics.

CORE INSTRUCTION

Warm-Up

- (5 min.) Have students review **"Tres mitos latinoamericanos,"** pp. 320–327.

Lectura

Assess

- (30 min.) Give **Prueba de lectura: "Tres mitos latinoamericanos,"** *Assessment Program,* pp. 167–168.

Vocabulario

Teach

- (10 min.) Have students do **Vocabulario en contexto,** Activity A, p. 337.

Wrap-Up

- (5 min.) Have students do Activity A, *TRB,* p. 176.

OPTIONAL RESOURCES

- (15 min.) Have students read **A leer por tu cuenta: "El corrido de Gregorio Cortez,"** pp. 333–335. See **"El corrido de Gregorio Cortez,"** *TRB,* pp. 173–174. ▲
- (15 min.) Have students listen to Audio CD 3, Track 10. ▲
- Have students practice the Reading Strategy, *TRB,* pp. 272–273. ◆ ●

Practice Options/Homework Suggestions

- *Cuaderno de práctica,* pp. 93–95
- Have students study **Vocabulario adicional, "El corrido de Gregorio Cortez,"** *TRB,* p. 301.
- *Cuaderno de práctica,* **Vocabulario adicional,** p. 149

▲ = Advanced Learners ◆ = Slower Pace Learners ● = Special Learning Needs ■ = Heritage Speakers

Teacher's Name ______________________ Class ______________________ Date ______________

COLECCIÓN 5

Mitos

DAY 13 50-MINUTE LESSON PLAN

STANDARDS FOR FOREIGN LANGUAGE LEARNING: DAY 13

Vocabulario

Communication 1.1: Students engage in conversations, provide and obtain information, express feelings and emotions, and exchange opinions.

Communication 1.2: Students understand and interpret written and spoken language on a variety of topics.

Communication 1.3: Students present information, concepts, and ideas to an audience of listeners or readers on a variety of topics.

Communities 5.1: Students use the language both within and beyond the school setting.

CORE INSTRUCTION

Warm-Up

- (5 min.) Have students review vocabulary for **"Tres mitos latinoamericanos,"** p. 363.

Vocabulario

Teach

- (10 min.) Have students do Activity C, p. 338.
- (10 min.) Have students do Activity D, pp. 338–339.
- (15 min.) Have students do Activity E, p. 339. You may wish to use Audio CD 2, Track 15. See script, *TRB,* p. 177.

Wrap-Up

- (10 min.) Have students do Pair Work, *TRB,* p. 177.

OPTIONAL RESOURCES

- (20 min.) Have students do Additional Practice, *TRB,* p. 176. ◆ ●
- (20 min.) Have pairs of students do Challenge, *TRB,* p. 177. ▲
- (10 min.) Have students do **Para hispanohablantes,** *TRB,* p. 170. ■
- (10 min.) Have students do **Para angloparlantes,** *TRB,* p. 170.

Practice Options/Homework Suggestions

- *Cuaderno de práctica,* Activities 1–2, pp. 96–97

Assessment Options

- *Assessment Program,* **Prueba de lectura: "El corrido de Gregorio Cortez,"** pp. 169–170
- *Assessment Program,* **Prueba de vocabulario adicional,** p. 271

▲ = Advanced Learners ◆ = Slower Pace Learners ● = Special Learning Needs ■ = Heritage Speakers

Teacher's Name ______________________ Class ____________________ Date __________

COLECCIÓN 5

Mitos

DAY 14 50-MINUTE LESSON PLAN

STANDARDS FOR FOREIGN LANGUAGE LEARNING: DAY 14

Vocabulario

Communication 1.1: Students engage in conversations, provide and obtain information, express feelings and emotions, and exchange opinions.

Communication 1.2: Students understand and interpret written and spoken language on a variety of topics.

Connections 3.1: Students reinforce and further their knowledge of other disciplines through the foreign language.

Comparisons 4.1: Students demonstrate understanding of the nature of language through comparisons of the language studied and their own.

CORE INSTRUCTION

Warm-Up

- (5 min.) Have students study the words in the chart, **Mejora tu vocabulario: Las voces indígenas,** p. 340.

Vocabulario

Teach

- (15 min.) Present **Mejora tu vocabulario: Las voces indígenas,** pp. 339–340. See **Mejora tu vocabulario: Las voces indígenas,** *TRB,* p. 177.
- (10 min.) Have students do Activity F, p. 340.
- (5 min.) Have students do Activity G, p. 340.
- (10 min.) Have students do Activity H, p. 341 (the first four words).

Wrap-Up

- (5 min.) Do the last five words of Activity H, p. 341. Have the class guess the words.

OPTIONAL RESOURCES

- (10 min.) Have students do Critical Thinking, *TRB,* p. 177. ▲
- (30 min.) Have students do Extension, *TRB,* pp. 177–178. ▲

Practice Options/Homework Suggestions

- *Cuaderno de práctica,* Activity 3, p. 97
- *Advanced Placement Literature Preparation Book,* pp. 109–117 ▲

▲ = Advanced Learners ◆ = Slower Pace Learners ● = Special Learning Needs ■ = Heritage Speakers

Mitos

DAY 15 50-MINUTE LESSON PLAN

STANDARDS FOR FOREIGN LANGUAGE LEARNING: DAY 15

Vocabulario

Communication 1.2: Students understand and interpret written and spoken language on a variety of topics.

Communication 1.3: Students present information, concepts, and ideas to an audience of listeners or readers on a variety of topics.

Comparisons 4.1: Students demonstrate understanding of the nature of language through comparisons of the language studied and their own.

Communities 5.1: Students use the language both within and beyond the school setting.

CORE INSTRUCTION

Warm-Up

- (5 min.) Have students begin Activity I, p. 341. See Activity I, *TRB,* p. 178.

Vocabulario

Teach

- (25 min.) Have students continue Activity I, p. 341. See Activity I, *TRB,* p. 178.
- (10 min.) Have students do Activity J, p. 341. See Activity J, *TRB,* p. 178.

Wrap-Up

- (10 min.) Have students do the second Extension, *TRB,* p. 178, as a class.

OPTIONAL RESOURCES

- (20 min.) Have students do **Para hispanohablantes,** *TRB,* p. 178. ■
- (10 min.) Have students do **Para angloparlantes,** *TRB,* p. 178.
- (20 min.) Have students do Group Work, *TRB,* p. 176.

Practice Options/Homework Suggestions

- Have students finish writing their creation myths for Activity I, p. 341.
- Have students study **Vocabulario esencial, Mejora tu vocabulario,** p. 363. ◆ ●
- Have students study for **Prueba de vocabulario.**

▲ = Advanced Learners ◆ = Slower Pace Learners ● = Special Learning Needs ■ = Heritage Speakers

Teacher's Name ______________ Class ______________ Date ______________

COLECCIÓN 5

Mitos

DAY 16 50-MINUTE LESSON PLAN

STANDARDS FOR FOREIGN LANGUAGE LEARNING: DAY 16

Vocabulario/Gramática

Communication 1.2: Students understand and interpret written and spoken language on a variety of topics.

Connections 3.1: Students reinforce and further their knowledge of other disciplines through the foreign language.

Comparisons 4.1: Students demonstrate understanding of the nature of language through comparisons of the language studied and their own.

CORE INSTRUCTION

Warm-Up

- (5 min.) Have students review **Vocabulario,** pp. 337–341.

Vocabulario

Assess

- (30 min.) Give **Prueba de vocabulario,** *Assessment Program,* pp. 171–172.

Gramática

Teach

- (10 min.) Present **El subjuntivo en las cláusulas adverbiales,** p. 347. See **El subjuntivo en las cláusulas adverbiales,** *TRB,* pp. 180–181.

Wrap-Up

- (5 min.) Have students review how to conjugate verbs in the present and past subjunctive.

OPTIONAL RESOURCES

- (20 min.) Have students do **Para angloparlantes,** *TRB,* p. 182.
- (15 min.) Have students do Pair Work, *TRB,* p. 181.

Practice Options/Homework Suggestions

- Have students study **El subjuntivo en las cláusulas adverbiales,** p. 347.
- *Cuaderno de práctica,* Activity 11, p. 103

▲ = Advanced Learners ◆ = Slower Pace Learners ● = Special Learning Needs ■ = Heritage Speakers

Teacher's Name ______________________ Class ____________________ Date ____________

COLECCIÓN 5

Mitos

DAY 17 50-MINUTE LESSON PLAN

STANDARDS FOR FOREIGN LANGUAGE LEARNING: DAY 17

Gramática

Communication 1.2: Students understand and interpret written and spoken language on a variety of topics.

Comparisons 4.1: Students demonstrate understanding of the nature of language through comparisons of the language studied and their own.

Communities 5.1: Students use the language both within and beyond the school setting.

Communities 5.2: Students show evidence of becoming lifelong learners by using the language for personal enjoyment and enrichment.

CORE INSTRUCTION

Warm-Up

- (5 min.) Have students review **El modo subjuntivo en cláusulas adverbiales, *Guía del lenguaje,*** p. R55.

Gramática

Teach

- (10 min.) Have students do Activity G, p. 348.
- (15 min.) Have students do Activity H, pp. 348–349.
- (15 min.) Have students do Activity I, p. 349.

Wrap-Up

- (5 min.) Have students review **Las cláusulas adverbiales y la secuencia de los tiempos,** p. 347.

OPTIONAL RESOURCES

- (30 min.) Have students do Activity J, p. 350.
- (20 min.) Have students do **Para hispanohablantes,** *TRB,* p. 182. ■
- (20 min.) Have students do **Ampliación, Hoja de práctica 5-A: Más sobre el subjuntivo en cláusulas adverbiales,** *TRB,* p. 289. ◆ ●

Practice Options/Homework Suggestions

- *Cuaderno de práctica,* Activities 12–15, pp. 104–105
- *Cuaderno de práctica,* **Ampliación, Hoja de práctica 5-A,** p. 150 ◆ ●
- Have students study for **Prueba de gramática.**
- *Advanced Placement Literature Preparation Book,* pp. 118–122 ▲

▲ = Advanced Learners ◆ = Slower Pace Learners ● = Special Learning Needs ■ = Heritage Speakers

Mitos

DAY 18 50-MINUTE LESSON PLAN

STANDARDS FOR FOREIGN LANGUAGE LEARNING: DAY 18

Gramática

Communication 1.2: Students understand and interpret written and spoken language on a variety of topics.

Comparisons 4.1: Students demonstrate understanding of the nature of language through comparisons of the language studied and their own.

CORE INSTRUCTION

Warm-Up

- (5 min.) Have students review **El subjuntivo en las cláusulas adverbiales,** p. 347.

Gramática
Assess

- (15 min.) Give **Prueba de gramática: El subjuntivo en las cláusulas adverbiales,** *Assessment Program,* p. 175.

Gramática
Teach

- (10 min.) Present **Comparación y contraste,** p. 350. See **Comparación y contraste,** *TRB,* p. 181.
- (10 min.) Have students do Activity A, pp. 350–351.

Wrap-Up

- (10 min.) Have students do Activity C, p. 351. See Activity C, *TRB,* p. 182.

OPTIONAL RESOURCES

- (10 min.) Have students do Activity B, p. 351.

Practice Options/Homework Suggestions

- *Cuaderno de práctica,* Activities 16–18, pp. 106–107
- Have students study for **Prueba de comparación y contraste.**

Assessment Options

- *Assessment Program:* **Prueba de gramática 5-A: Más sobre el subjuntivo en cláusulas adverbiales,** p. 272

▲ = Advanced Learners ◆ = Slower Pace Learners ● = Special Learning Needs ■ = Heritage Speakers

Teacher's Name ______________________ Class ______________________ Date __________

COLECCIÓN

Mitos

DAY 19 50-MINUTE LESSON PLAN

STANDARDS FOR FOREIGN LANGUAGE LEARNING: DAY 19

Gramática/Panorama cultural

Communication 1.2: Students understand and interpret written and spoken language on a variety of topics.

Connections 3.1: Students reinforce and further their knowledge of other disciplines through the foreign language.

CORE INSTRUCTION

Warm-Up

- (5 min.) Have students review **Comparación y contraste,** pp. 350–351.

Gramática

Assess

- (20 min.) Give **Prueba de comparación y contraste,** *Assessment Program,* p. 176.

Panorama cultural

Teach

- (15 min.) Have students discuss the introduction of **Panorama cultural,** p. 330. See Summary and Presentation, **Panorama cultural,** *TRB,* p. 171.

Wrap-Up

- (10 min.) Have students do **Panorama cultural,** Teaching Suggestions, **Antes de ver,** item one, *Video Guide,* p. 40.

OPTIONAL RESOURCES

- (15 min.) Play Audio CD 2, Tracks 11–13, and have students listen to the interviews.

Practice Options/Homework Suggestions

- Have students read **Panorama cultural,** pp. 330–331.
- Have students do Family Link, *TRB,* p. 171. ▲ ■

▲ = Advanced Learners ◆ = Slower Pace Learners ● = Special Learning Needs ■ = Heritage Speakers

Teacher's Name ______________________ Class ______________________ Date ____________

COLECCIÓN 5

Mitos

DAY 20 50-MINUTE LESSON PLAN

STANDARDS FOR FOREIGN LANGUAGE LEARNING: DAY 20

Panorama cultural

Communication 1.1: Students engage in conversations, provide and obtain information, express feelings and emotions, and exchange opinions.

Communication 1.2: Students understand and interpret written and spoken language on a variety of topics.

Connections 3.2: Students acquire information and recognize the distinctive viewpoints that are only available through the foreign language and its cultures.

CORE INSTRUCTION

Warm-Up

- (5 min.) Have students do **Hoja de actividades 3, Antes de ver,** *Video Guide,* p. 44.

Panorama cultural

Teach

- (20 min.) Show **Panorama cultural,** *Video Program* (Videocassette 2). See **Panorama cultural,** Teaching Suggestions, **Mientras lo ves,** *Video Guide,* p. 40.
- (15 min.) Have students do **Hoja de actividades 3, Mientras lo ves,** *Video Guide,* p. 44.

Wrap-Up

- (10 min.) Have students do **Hoja de actividades 3, Después de ver,** *Video Guide,* p. 44.

OPTIONAL RESOURCES

- (20 min.) Have students do **Panorama cultural,** Teaching Suggestions, **Después de ver,** item one, *Video Guide,* p. 40.
- (20 min.) Have students do Activity D, p. 331.
- (10 min.) Have students do Activity E, p. 331. You may use Audio CD 2, Track 14. See script, *TRB,* p. 171.

Practice Options/Homework Suggestions

- Have students do **Para pensar y hablar,** Activities A–C, p. 331.
- *Advanced Placement Literature Preparation Book,* pp. 138–146 ▲

▲ = Advanced Learners ◆ = Slower Pace Learners ● = Special Learning Needs ■ = Heritage Speakers

Teacher's Name ______________________ Class ______________________ Date ______________

Mitos

DAY 21 50-MINUTE LESSON PLAN

STANDARDS FOR FOREIGN LANGUAGE LEARNING: DAY 21

Comunidad y oficio

Communication 1.1: Students engage in conversations, provide and obtain information, express feelings and emotions, and exchange opinions.

Communication 1.2: Students understand and interpret written and spoken language on a variety of topics.

Communication 1.3: Students present information, concepts, and ideas to an audience of listeners or readers on a variety of topics.

Connections 3.1: Students reinforce and further their knowledge of other disciplines through the foreign language.

Communities 5.1: Students use the language both within and beyond the school setting.

Communities 5.2: Students show evidence of becoming lifelong learners by using the language for personal enjoyment and enrichment.

CORE INSTRUCTION

Warm-Up

- (5 min.) Have students answer the questions in Getting Started, *TRB,* p. 175.

Comunidad y oficio

Teach

- (10 min.) Have students read **Comunidad y oficio,** p. 336.
- (10 min.) Have students do **Comunidad y oficio,** Teaching Suggestions, **Antes de ver,** *Video Guide,* p. 41.
- (15 min.) Show **Comunidad y oficio,** *Video Program* (Videocassette 2). See **Comunidad y oficio,** Teaching Suggestions, **Mientras lo ves,** *Video Guide,* p. 41.
- (5 min.) Have students do **Hoja de actividades 4, Mientras lo ves,** *Video Guide,* p. 45.

Wrap-Up

- (5 min.) Have students do **Hoja de actividades 4, Después de ver,** item two, *Video Guide,* p. 45.

OPTIONAL RESOURCES

- (5 min.) Have students do **Comunidad y oficio,** Teaching Suggestions, **Después de ver,** *Video Guide,* p. 41.
- (15 min.) Have students do **Hoja de actividades 4, Después de ver,** item one, *Video Guide,* p. 45.
- (30 min.) Have students do **Investigaciones,** Activity A or B, p. 336.
- (30 min.) Have students do Science Link, *TRB,* p. 175. ▲
- (15 min.) Have students do Thinking Critically, *TRB,* p. 175. ▲
- (15 min.) Have students listen to Tish Hinojosa recordings and discuss, p. 336.

Practice Options/Homework Suggestions

- Internet (go.hrw.com, Keyword: WN3 MITOS-CYO)
- Have students do History Link, *TRB,* p. 175.

▲ = Advanced Learners ◆ = Slower Pace Learners ● = Special Learning Needs ■ = Heritage Speakers

Teacher's Name ______________________ Class ______________________ Date ____________

COLECCIÓN 5

Mitos

DAY 22 50-MINUTE LESSON PLAN

STANDARDS FOR FOREIGN LANGUAGE LEARNING: DAY 22

Ortografía

Communication 1.2: Students understand and interpret written and spoken language on a variety of topics.

Comparisons 4.1: Students demonstrate understanding of the nature of language through comparisons of the language studied and their own.

CORE INSTRUCTION

Warm-Up

- (5 min.) Have students read the introduction of **El sonido /s/,** p. 352.

Ortografía

Teach

- (20 min.) Present **El sonido /s/,** p. 352. See **El sonido /s/,** *TRB,* p. 183.
- (10 min.) Have students do Activity A, p. 353. See Activity A, *TRB,* p. 183.
- (10 min.) Have students do Activity B, p. 353.

Wrap-Up

- (5 min.) Have students review **¡Ojos!,** p. 352.

OPTIONAL RESOURCES

- (10 min.) See Language Note, *TRB,* p. 183.
- (30 min.) Have students do Group Work, *TRB,* pp. 183–184. ▲
- (10 min.) Have students do the first Additional Practice, *TRB,* p. 184. ◆ ●
- (20 min.) Have students do Activity C, p. 353.
- (20 min.) Have students do **Ampliación, Hoja de práctica 5-B: La aspiración,** *TRB,* p. 290. ◆ ●

Practice Options/Homework Suggestions

- Have students do Activity D, p. 354.
- *Cuaderno de práctica,* Activities 1–3, p. 108
- *Cuaderno de práctica,* **Ampliación, Hoja de práctica 5-B,** p. 151 ◆ ●

▲ = Advanced Learners ◆ = Slower Pace Learners ● = Special Learning Needs ■ = Heritage Speakers

Teacher's Name ______________________ Class ____________________ Date __________

COLECCIÓN

Mitos

DAY 23 50-MINUTE LESSON PLAN

STANDARDS FOR FOREIGN LANGUAGE LEARNING: DAY 23

Ortografía

Communication 1.1: Students engage in conversations, provide and obtain information, express feelings and emotions, and exchange opinions.

Communication 1.2: Students understand and interpret written and spoken language on a variety of topics.

Comparisons 4.1: Students demonstrate understanding of the nature of language through comparisons of the language studied and their own.

CORE INSTRUCTION

Warm-Up

- (5 min.) Have students review **El sonido /s/,** pp. 352–353.

Ortografía

Teach

- (5 min.) Present **La acentuación,** pp. 354. See **La acentuación,** *TRB,* p. 184.
- (15 min.) Have students do Activities F and G, p. 355.
- (20 min.) Give **Dictado,** Activities A and B, p. 355. You may wish to use Audio CD 2, Tracks 16. See scripts, *TRB,* p. 185.

Wrap-Up

- (5 min.) Review rules for accent marks with students. See **El acento,** pp. R64–R67.

OPTIONAL RESOURCES

- (15 min.) Have students do **La acentuación: Formas verbales,** *TRB,* p. 184.
- (15 min.) Have students do Activity E, p. 354. See Activity E, *TRB,* p. 184.
- (15 min.) Have students do second Additional Practice or Cooperative Learning, *TRB,* pp. 184–185. ◆ ●
- (15 min.) Have students do Pair Work, *TRB,* p. 185.
- (20 min.) Have students do **Para hispanohablantes,** *TRB,* p. 185. ■
- (20 min.) Have students do **Para angloparlantes,** *TRB,* p. 165.

Practice Options/Homework Suggestions

- *Cuaderno de práctica,* Activities 4–5, p. 109
- Have students study for **Prueba de ortografía.**
- *Advanced Placement Literature Preparation Book,* pp. 147–153 ▲

Assessment Options

- *Assessment Program,* **Prueba de ortografía 5-B: La aspiración,** p. 273

▲ = Advanced Learners ◆ = Slower Pace Learners ● = Special Learning Needs ■ = Heritage Speakers

Teacher's Name ______________ Class ______________ Date ______________

COLECCIÓN 5

Mitos

DAY 24 50-MINUTE LESSON PLAN

STANDARDS FOR FOREIGN LANGUAGE LEARNING: DAY 24

Ortografía/Taller del escritor

Communication 1.2: Students understand and interpret written and spoken language on a variety of topics.

Communication 1.3: Students present information, concepts, and ideas to an audience of listeners or readers on a variety of topics.

CORE INSTRUCTION

Warm-Up

- (5 min.) Have students review **Ortografía,** pp. 352–355.

Ortografía

Assess

- (20 min.) Give **Prueba de ortografía,** *Assessment Program,* p. 178.

Taller del escritor

Teach

- (5 min.) Introduce **Taller del escritor,** p. 356. See Presenting the Workshop, *TRB,* p. 186.
- (10 min.) Present **Antes de escribir,** *TRB,* p. 186.

Wrap-Up

- (10 min.) Have students do **Cuaderno del escritor,** p. 356.

OPTIONAL RESOURCES

- (20 min.) Have students read and discuss **Esquema para una evaluación,** p. 357.

Practice Options/Homework Suggestions

- Have students choose the theme of their evaluation essay, p. 357.

▲ = Advanced Learners ◆ = Slower Pace Learners ● = Special Learning Needs ■ = Heritage Speakers

Teacher's Name ______________ Class ______________ Date ______________

COLECCIÓN 5

Mitos

DAY 25 50-MINUTE LESSON PLAN

STANDARDS FOR FOREIGN LANGUAGE LEARNING: DAY 25

Taller del escritor

Communication 1.2: Students understand and interpret written and spoken language on a variety of topics.

Communication 1.3: Students present information, concepts, and ideas to an audience of listeners or readers on a variety of topics.

Connections 3.1: Students reinforce and further their knowledge of other disciplines through the foreign language.

Communities 5.1: Students use the language both within and beyond the school setting.

CORE INSTRUCTION

Warm-Up

- (5 min.) Have students study the chart in **Prepara un cuadro,** p. 356.

Taller del escritor

Teach

- (20 min.) In groups or individually, have students do a chart for their own essay, p. 356.
- (20 min.) Have students do **Establece criterios para emitir un juicio,** p. 357.

Wrap-Up

- (5 min.) Have students review the charts and compare them to the chart modeled on p. 356.

OPTIONAL RESOURCES

- (15 min.) Have students read reviews of movies, books, or concerts in Spanish-language newspapers and magazines.

Practice Options/Homework Suggestions

- Have students read **El borrador** and **Desarrolla tu propio estilo: Connotaciones,** pp. 357–358.
- Have students begin writing their evaluations.

▲ = Advanced Learners ◆ = Slower Pace Learners ● = Special Learning Needs ■ = Heritage Speakers

Teacher's Name ____________________ Class ____________________ Date ____________

Mitos

DAY 26 50-MINUTE LESSON PLAN

STANDARDS FOR FOREIGN LANGUAGE LEARNING: DAY 26

Taller del escritor

Communication 1.1: Students engage in conversations, provide and obtain information, express feelings and emotions, and exchange opinions.

Communication 1.2: Students understand and interpret written and spoken language on a variety of topics.

Communication 1.3: Students present information, concepts, and ideas to an audience of listeners or readers on a variety of topics.

Communities 5.1: Students use the language both within and beyond the school setting.

CORE INSTRUCTION

Warm-Up

- (5 min.) Have students review the two methods for presenting information in **El borrador,** p. 357, and **Pautas para redactar,** p. 358.

Taller del escritor

Teach

- (30 min.) Have students work on their first draft of **Una evaluación.**
- (10 min.) Have students do **Intercambio entre compañeros,** p. 358. Have them use the expressions in **Así se dice,** p. 358, to evaluate their partner's work.

Wrap-Up

- (5 min.) Have student volunteers give examples of the feedback they gave their partner.

OPTIONAL RESOURCES

- (20 min.) Have students do Reteaching, *TRB,* p. 187. ◆ ●

Practice Options/Homework Suggestion

- Have students revise their drafts using the suggestions in **Autoevaluación,** item two, p. 359.

▲ = Advanced Learners ◆ = Slower Pace Learners ● = Special Learning Needs ■ = Heritage Speakers

Teacher's Name ______________________ Class ______________________ Date __________

Mitos

DAY 27 50-MINUTE LESSON PLAN

STANDARDS FOR FOREIGN LANGUAGE LEARNING: DAY 27

Taller del escritor

Communication 1.2: Students understand and interpret written and spoken language on a variety of topics.

Communication 1.3: Students present information, concepts, and ideas to an audience of listeners or readers on a variety of topics.

CORE INSTRUCTION

Warm-Up

- (5 min.) Have students read the first **Modelo,** p. 359.

Taller del escritor

Teach

- (10 min.) Have students read the second **Modelo,** p. 360.
- (20 min.) Have students do **Corrección de pruebas,** p. 360. See **Corrección de pruebas,** *TRB,* p. 187.
- (10 min.) Have students do **Reflexión,** p. 360. Have them use the expressions in **Así se dice,** p. 360. See **Reflexión,** *TRB,* p. 187.

Wrap-Up

- (5 min.) Have students do Closure, *TRB,* p. 187.

OPTIONAL RESOURCES

- (15 min.) Have students do one of the activities in **Publicación,** p. 360.
- (20 min.) Have students do **Publicación,** *TRB,* p. 187. ▲

Practice Options/Homework Suggestions

- Have students do **A ver si puedo…,** pp. 361–362.

▲ = Advanced Learners ◆ = Slower Pace Learners ● = Special Learning Needs ■ = Heritage Speakers

Mitos

DAY 28 50-MINUTE LESSON PLAN

STANDARDS FOR FOREIGN LANGUAGE LEARNING: DAY 28

A ver si puedo...

Communication 1.2: Students understand and interpret written and spoken language on a variety of topics.

Cultures 2.1: Students demonstrate an understanding of the relationship between the practices and perspectives of the culture studied.

Cultures 2.2: Students demonstrate an understanding of the relationship between the products and perspectives of the culture studied.

Connections 3.1: Students reinforce and further their knowledge of other disciplines through the foreign language.

Connections 3.2: Students acquire information and recognize the distinctive viewpoints that are only available through the foreign language and its cultures.

CORE INSTRUCTION

Warm-Up

- (5 min.) Have students review the objectives listed on the Collection Opener, p. 302.

A ver si puedo...

Review

- (10 min.) Have students do **Lectura,** Activities A and B, p. 361.
- (10 min.) Have students do **Cultura,** Activity C, p. 361.
- (10 min.) Have students do **Comunicación,** Activities D, E, F, and G, pp. 361–362.
- (10 min.) Have students do **Escritura,** Activities H, I, and J, p. 362.

Wrap-Up

- (5 min.) Answer any questions about either of the two chapter exams.

OPTIONAL RESOURCES

- (35 min.) Have students read **Enlaces literarios: La nueva narrativa latinoamericana del siglo XX,** pp. 364–374. See *TRB,* pp. 190–193. ▲
- (20 min.) Have students do **Comprensión del texto,** p. 375. See **Comprensión del texto,** *TRB,* p. 192. ▲
- (25 min.) Have students do **Análisis del texto,** p. 375. See **Análisis del texto,** *TRB,* pp. 192–193. ▲
- (25 min.) Have students do **Más allá del texto,** p. 375. See **Más allá del texto,** *TRB,* p. 193. ▲

Practice Options/Homework Suggestions

- Have students study for the **Examen de lengua.**

▲ = Advanced Learners ◆ = Slower Pace Learners ● = Special Learning Needs ■ = Heritage Speakers

Teacher's Name ______________________ Class ______________________ Date ____________

COLECCIÓN **5**

Mitos

DAY 29 50-MINUTE LESSON PLAN

CORE INSTRUCTION

Assess

- (50 min.) Give **Colección 5 Examen de lengua,** *Assessment Program,* pp. 185–192.

OPTIONAL RESOURCES

- (50 min.) Give **Examen de lectura: del *Popul Vuh* y "Tres mitos latinoamericanos,"** *Assessment Program,* pp. 179–183. To allow students more time to take the exam, either **Examen** may be given over two class periods.

Practice Options/Homework Suggestions

- Have students study for the **Examen de lectura: del *Popul Vuh,* y "Tres mitos latinoamericanos."**

Assessment Options

- *Assessment Program,* Performance Assessment, p. 299

▲ = Advanced Learners ◆ = Slower Pace Learners ● = Special Learning Needs ■ = Heritage Speakers

Mitos

CORE INSTRUCTION

Assess

- (50 min.) Give **Examen de lectura: del *Popul Vuh* y "Tres mitos latinoamericanos,"** *Assessment Program,* pp. 179–183.

OPTIONAL RESOURCES

- (50 min.) Give **Colección 5 Examen de lengua,** *Assessment Program,* pp. 185–192. To allow students more time to take the exam, either **Examen** may be given over two class periods.

Practice Options/Homework Suggestions

- Internet (go.hrw.com, Keyword: WN3 MITOS)

Assessment Options

- *Assessment Program,* Performance Assessment, p. 299.

▲ = Advanced Learners ◆ = Slower Pace Learners ● = Special Learning Needs ■ = Heritage Speakers

Teacher's Name ______________________ Class ____________________ Date __________

COLECCIÓN 6

Perspectivas humorísticas

DAY 1 50-MINUTE LESSON PLAN

STANDARDS FOR FOREIGN LANGUAGE LEARNING: DAY 1

Lectura

Communication 1.1: Students engage in conversations, provide and obtain information, express feelings and emotions, and exchange opinions.

Communication 1.2: Students understand and interpret written and spoken language on a variety of topics.

Connections 3.1: Students reinforce and further their knowledge of other disciplines through the foreign language.

CORE INSTRUCTION

Warm-Up

- (5 min.) Have students read the objectives on the Collection Opener, p. 376. See Collection Overview, *TRB,* p. 198.

Lectura

Teach

- (15 min.) Have students read **Punto de partida,** p. 378. Have students do **Punto de partida,** *TRB,* p. 199.
- (5 min.) Have students read **Elementos de literatura,** p. 378.
- (20 min.) Have students do **Comparte tus ideas,** p. 378. See **Comparte tus ideas,** *TRB,* p. 199.

Wrap-Up

- (5 min.) Have students do **Elementos de literatura,** *TRB,* p. 199.

OPTIONAL RESOURCES

- (10 min.) See Presentation Suggestions, *TRB,* p. 198.
- (10 min.) Read aloud **de *Don Quijote de la Mancha,*** Summary, *TRB,* p. 199. ◆ ●

Practice Options/Homework Suggestions

- Internet (go.hrw.com, Keyword: WN3 PERSPECTIVAS-LEC)
- Have students read **de *Don Quijote de la Mancha,*** pp. 380–382.
- Have students study **Vocabulario esencial, de *Don Quijote de la Mancha,*** p. 427. ◆ ●

▲ = Advanced Learners ◆ = Slower Pace Learners ● = Special Learning Needs ■ = Heritage Speakers

Teacher's Name ______________________ Class ______________________ Date ____________

COLECCIÓN 6

Perspectivas humorísticas

DAY 2 50-MINUTE LESSON PLAN

STANDARDS FOR FOREIGN LANGUAGE LEARNING: DAY 2

Lectura/Vocabulario

Communication 1.2: Students understand and interpret written and spoken language on a variety of topics.

Connections 3.1: Students reinforce and further their knowledge of other disciplines through the foreign language.

Connections 3.2: Students acquire information and recognize the distinctive viewpoints that are only available through the foreign language and its cultures.

CORE INSTRUCTION

Warm-Up

- (5 min.) Have students read the definition of **parodia** on p. 378, and in ***Glosario de términos literarios,*** p. R11.

Lectura

Teach

- (20 min.) Have students read aloud **de *Don Quijote de la Mancha,*** pp. 380–382.

Vocabulario

Teach

- (15 min.) In small groups, have students do **Repaso,** p. 384. See **Repaso,** *TRB,* p. 200.

Wrap-Up

- (10 min.) Have students read aloud **Conoce al escritor,** p. 383.

OPTIONAL RESOURCES

- (20 min.) See Techniques for Handling the Reading, *TRB,* p. 199. ◆ ●
- Have students practice the Reading Strategy, *TRB,* p. 274. ◆ ●
- (10 min.) Have students listen to the recording of **de *Don Quijote de la Mancha.*** You may use Audio CD 3, Track 11.

Practice Options/Homework Suggestions

- Internet (go.hrw.com, Keyword: WN3 PERSPECTIVAS-LEC)
- Have students study **Vocabulario adicional, de *Don Quijote de la Mancha,*** *TRB,* p. 302. ◆ ●
- *Cuaderno de práctica,* Activities 1–4, pp. 111–112

▲ = Advanced Learners ◆ = Slower Pace Learners ● = Special Learning Needs ■ = Heritage Speakers

Teacher's Name ______________________ Class ______________________ Date ____________

COLECCIÓN 6

Perspectivas humorísticas

DAY 3 50-MINUTE LESSON PLAN

STANDARDS FOR FOREIGN LANGUAGE LEARNING: DAY 3

Lectura

Communication 1.1: Students engage in conversations, provide and obtain information, express feelings and emotions, and exchange opinions.

Communication 1.2: Students understand and interpret written and spoken language on a variety of topics.

Comparisons 4.1: Students demonstrate understanding of the nature of language through comparisons of the language studied and their own.

Comparisons 4.2: Students demonstrate understanding of the concept of culture through comparisons of the cultures studied and their own.

CORE INSTRUCTION

Warm-Up

- (5 min.) In pairs, have students do **Primeras impresiones,** p. 384.

Lectura

Teach

- (20 min.) Have students do **Interpretaciones del texto,** p. 384. See **Interpretaciones del texto,** *TRB,* p. 200.
- (15 min.) Present **Así se dice** and **¿Te acuerdas?,** p. 384.

Wrap-Up

- (10 min.) Have students answer questions in **Conexiones con el texto,** p. 384, using the expressions in **Así se dice.**

OPTIONAL RESOURCES

- (20 min.) Have students do **Más allá del texto,** p. 384. See **Más allá del texto,** *TRB,* p. 200.
- (20 min.) Have students do **Cuaderno del escritor,** p. 385. See **Cuaderno del escritor,** *TRB,* p. 200. Have students review **Así se dice** and **¿Te acuerdas?,** p. 384.
- (30 min.) Have students do **Escritura y dibujo creativos** or **Arte,** p. 385. See **Escritura y dibujo creativos** or **Arte,** *TRB,* p. 200.
- (20 min.) Have students do **Para hispanohablantes,** *TRB,* p. 201. ■
- (20 min.) Have students do **Para angloparlantes,** *TRB,* p. 201.

Practice Options/Homework Suggestions

- Internet (go.hrw.com, Keyword: WN3 PERSPECTIVAS-LEC)
- *Cuaderno de práctica,* p. 118
- Have students study for **Prueba de lectura.**

▲ = Advanced Learners ◆ = Slower Pace Learners ● = Special Learning Needs ■ = Heritage Speakers

Teacher's Name ______________________ Class ______________________ Date __________

COLECCIÓN 6

Perspectivas humorísticas

DAY 4 50-MINUTE LESSON PLAN

STANDARDS FOR FOREIGN LANGUAGE LEARNING: DAY 4

Vocabulario/Lectura

Communication 1.1: Students engage in conversations, provide and obtain information, express feelings and emotions, and exchange opinions.

Communication 1.2: Students understand and interpret written and spoken language on a variety of topics.

Connections 3.1: Students reinforce and further their knowledge of other disciplines through the foreign language.

CORE INSTRUCTION

Warm-Up

- (5 min.) Have students review **de *Don Quijote de la Mancha,*** pp. 380–382.

Vocabulario

Teach

- (10 min.) Have students do **Vocabulario en contexto,** Activities A and B, p. 405.

Lectura

Assess

- (30 min.) Give **Prueba de lectura: de *Don Quijote de la Mancha,*** *Assessment Program,* pp. 199–200.

Wrap-Up

- (5 min.) Have students discuss the art on pp. 381 and 382.

OPTIONAL RESOURCES

- (10 min.) Show **Lectura: Fuenteovejuna,** *Video Program* (Videocassette 2). See Teaching Suggestions, *Video Guide,* p. 47, and **Hoja de actividades 1,** p. 51.

Practice Options/Homework Suggestions

- Have students study **Gramática: El aspecto: repaso y ampliación, El aspecto perfectivo,** and **El aspecto imperfectivo,** pp. 410–412.

▲ = Advanced Learners ◆ = Slower Pace Learners ● = Special Learning Needs ■ = Heritage Speakers

Teacher's Name ________________ Class ________________ Date ________

COLECCIÓN 6

Perspectivas humorísticas

DAY 5 50-MINUTE LESSON PLAN

STANDARDS FOR FOREIGN LANGUAGE LEARNING: DAY 5

Gramática

Communication 1.1: Students engage in conversations, provide and obtain information, express feelings and emotions, and exchange opinions.

Communication 1.2: Students understand and interpret written and spoken language on a variety of topics.

Comparisons 4.1: Students demonstrate understanding of the nature of language through comparisons of the language studied and their own.

CORE INSTRUCTION

Warm-Up

- (5 min.) Have students do Activity A, p. 410.

Gramática

Teach

- (15 min.) Present **El aspecto: repaso y ampliación,** p. 410. Review answers to Activity A, p. 410, with students. See **El aspecto: repaso y ampliación,** *TRB,* p. 217.
- (20 min.) Present **El aspecto perfectivo** and **El aspecto imperfectivo,** pp. 410–412. See **El aspecto perfectivo** and **El aspecto imperfectivo,** *TRB,* pp. 217–218.

Wrap-Up

- (10 min.) Have students do Activity B, p. 412. See Activity B, *TRB,* p. 218.

OPTIONAL RESOURCES

- (10 min.) Have students do Pair Work, *TRB,* p. 218.
- (10 min.) Review conjugations, both regular and irregular, of the preterite and imperfect with students.

Practice Options/Homework Suggestions

- *Cuaderno de práctica,* Activities 1–5, pp. 120–122
- Have students do Activity C, p. 412.
- Have students study for **Prueba de gramática.**
- *Advanced Placement Literature Preparation Book,* pp. 80–84 ▲

▲ = Advanced Learners ◆ = Slower Pace Learners ● = Special Learning Needs ■ = Heritage Speakers

Perspectivas humorísticas

DAY 6 50-MINUTE LESSON PLAN

STANDARDS FOR FOREIGN LANGUAGE LEARNING: DAY 6

Gramática/Cultura y lengua

Communication 1.2: Students understand and interpret written and spoken language on a variety of topics.

Communication 1.3: Students present information, concepts, and ideas to an audience of listeners or readers on a variety of topics.

Cultures 2.1: Students demonstrate an understanding of the relationship between the practices and perspectives of the culture studied.

Cultures 2.2: Students demonstrate an understanding of the relationship between the products and perspectives of the culture studied.

Connections 3.1: Students reinforce and further their knowledge of other disciplines through the foreign language.

CORE INSTRUCTION

Warm-Up

- (5 min.) Have students review **El aspecto perfectivo** and **El aspecto imperfectivo,** pp. 410–412.

Gramática

Assess

- (20 min.) Give **Prueba de gramática: El aspecto perfectivo e imperfectivo,** *Assessment Program,* p. 207.

Cultura y lengua

Teach

- (5 min.) Present **Cultura y lengua: España,** pp. 386–387. See Background Information, *TRB,* p. 202.
- (15 min.) Divide class into groups of five. Have each group read one of the five paragraphs on pp. 387–388 and write a two-sentence summary.

Wrap-Up

- (5 min.) Have one student from each group read their summary to the class.

OPTIONAL RESOURCES

- (15 min.) See **Cultura y lengua,** Teaching Suggestions, *Video Guide,* p. 48.
- (20 min.) Have students do Art Link, *TRB,* p. 202.
- (20 min.) Have students do History Link, *TRB,* pp. 202–203.

Practice Options/Homework Suggestions

- Internet (go.hrw.com, Keyword: WN3 PERSPECTIVAS-CYL)
- Have students read **El Siglo de Oro,** pp. 386–387.
- Have students do Project, *TRB,* p. 202.

▲ = Advanced Learners ◆ = Slower Pace Learners ● = Special Learning Needs ■ = Heritage Speakers

Teacher's Name ______________________ Class ______________________ Date ____________

COLECCIÓN 6

Perspectivas humorísticas

DAY 7 50-MINUTE LESSON PLAN

STANDARDS FOR FOREIGN LANGUAGE LEARNING: DAY 7

Cultura y lengua

Communication 1.1: Students engage in conversations, provide and obtain information, express feelings and emotions, and exchange opinions.

Communication 1.2: Students understand and interpret written and spoken language on a variety of topics.

Communication 1.3: Students present information, concepts, and ideas to an audience of listeners or readers on a variety of topics.

Connections 3.1: Students reinforce and further their knowledge of other disciplines through the foreign language.

Comparisons 4.1: Students demonstrate understanding of the nature of language through comparisons of the language studied and their own.

Communities 5.1: Students use the language both within and beyond the school setting.

Communities 5.2: Students show evidence of becoming lifelong learners by using the language for personal enjoyment and enrichment.

CORE INSTRUCTION

Warm-Up

- (5 min.) Have students discuss Project, *TRB,* p. 202.

Cultura y lengua

Teach

- (5 min.) Present **Así se dice,** p. 389. See **Así se dice,** *TRB,* p. 203.
- (25 min.) Have students read **La lengua del Siglo de Oro,** p. 389. See Thinking Critically, *TRB,* p. 202, and the first Thinking Critically, *TRB,* p. 203. See also Language Link, *TRB,* p. 203.
- (10 min.) Show **Cultura y lengua: El legato de Don Quijote de la Mancha,** *Video Program* (Videocassette 2). See **Cultura y lengua,** Teaching Suggestions, **Mientras lo ves,** *Video Guide,* p. 48.

Wrap-Up

- (5 min.) Have students listen to music from ***Man of la Mancha*** (the Broadway musical) and review photos on pp. 387–388.

OPTIONAL RESOURCES

- (20 min.) Have students do the second Thinking Critically, *TRB,* p. 203. ▲ ■
- (20 min.) Have students do Literature Link, *TRB,* p. 203. ▲
- (10 min.) Have students do **Hoja de actividades 2,** *Video Guide,* p. 52.

Practice Options/Homework Suggestions

- Internet (go.hrw.com, Keyword: WN3 PERSPECTIVAS-CYL)
- Have students do **Actividad,** p. 389.
- Have students study for **Prueba de cultura.**

▲ = Advanced Learners ◆ = Slower Pace Learners ● = Special Learning Needs ■ = Heritage Speakers

Teacher's Name ______________________ Class ______________________ Date ______________

COLECCIÓN 6

Perspectivas humorísticas

DAY 8 50-MINUTE LESSON PLAN

STANDARDS FOR FOREIGN LANGUAGE LEARNING: DAY 8

Cultura y lengua/Lectura

Communication 1.2: Students understand and interpret written and spoken language on a variety of topics.

Connections 3.1: Students reinforce and further their knowledge of other disciplines through the foreign language.

CORE INSTRUCTION

Warm-Up

- (5 min.) Have students review **Cultura y lengua,** pp. 386–389.

Cultura y lengua

Assess

- (20 min.) Give **Prueba de cultura: El Siglo de Oro,** *Assessment Program,* p. 211.

Lectura

Teach

- (15 min.) Have students read **Punto de partida,** p. 390. See **Punto de partida,** *TRB,* p. 204.
- (5 min.) Have students read **Elementos de literatura,** p. 399. See **Elementos de literatura,** *TRB,* p. 208.

Wrap-Up

- (5 min.) Have students give examples of surprise endings or funny plots they have read in a book or seen in a movie.

OPTIONAL RESOURCES

- Have students do Techniques for Handling the Reading, *TRB,* p. 204. ◆ ●

Practice Options/Homework Suggestions

- Internet (go.hrw.com, Keyword: WN3 PERSPECTIVAS-LEC)
- Have students read **Estrategias para leer,** p. 398. ◆ ●
- Have students read **"El libro talonario,"** pp. 391–395.

▲ = Advanced Learners ◆ = Slower Pace Learners ● = Special Learning Needs ■ = Heritage Speakers

Teacher's Name ______________________ Class ____________________ Date __________

COLECCIÓN 6

Perspectivas humorísticas

DAY 9 50-MINUTE LESSON PLAN

STANDARDS FOR FOREIGN LANGUAGE LEARNING: DAY 9

Lectura

Communication 1.1: Students engage in conversations, provide and obtain information, express feelings and emotions, and exchange opinions.

Communication 1.2: Students understand and interpret written and spoken language on a variety of topics.

Communication 1.3: Students present information, concepts, and ideas to an audience of listeners or readers on a variety of topics.

Connections 3.1: Students reinforce and further their knowledge of other disciplines through the foreign language.

Communities 5.1: Students use the language both within and beyond the school setting.

CORE INSTRUCTION

Warm-Up

- (5 min.) Have students locate Cádiz, Huelva, Sevilla, and Andalucía on a map.

Lectura

Teach

- (30 min.) Have students read aloud **"El libro talonario,"** pp. 391–395.
- (10 min.) Have students do **Repaso del texto,** p. 396. See **Repaso del texto,** *TRB,* p. 205.

Wrap-Up

- (5 min.) Have student volunteers share their character profiles with the class.

OPTIONAL RESOURCES

- (10 min.) Read aloud **"El libro talonario,"** Summary, *TRB,* p. 204. ◆ ●
- (20 min.) Have students do **Diálogo con el texto,** p. 390. See **Diálogo con el texto,** *TRB,* p. 204. ▲
- (30 min.) Have students practice the Reading Strategy, *TRB,* p. 275. ▲ ◆ ●

Practice Options/Homework Suggestions

- Internet (go.hrw.com, Keyword: WN3 PERSPECTIVAS-LEC)
- *Cuaderno de práctica,* Activities 1–4, pp. 113–114
- Have students study **Vocabulario esencial, "El libro talonario,"** p. 427. ◆ ●

▲ = Advanced Learners ◆ = Slower Pace Learners ● = Special Learning Needs ■ = Heritage Speakers

Teacher's Name ______________________ Class ______________________ Date ______________

COLECCIÓN 6

Perspectivas humorísticas

DAY 10 50-MINUTE LESSON PLAN

STANDARDS FOR FOREIGN LANGUAGE LEARNING: DAY 10

Lectura

Communication 1.2: Students understand and interpret written and spoken language on a variety of topics.

Connections 3.1: Students reinforce and further their knowledge of other disciplines through the foreign language.

Connections 3.2: Students acquire information and recognize the distinctive viewpoints that are only available through the foreign language and its cultures.

CORE INSTRUCTION

Warm-Up

- (5 min.) Have students read **Conoce al escritor,** p. 395.

Lectura

Teach

- (30 min.) Have students do **Interpretaciones del texto,** p. 396. See **Interpretaciones del texto,** *TRB,* p. 205.
- (10 min.) Have students do **Preguntas al texto,** item seven, p. 396.

Wrap-Up

- (5 min.) Have students do **Preguntas al texto,** item eight, p. 396.

OPTIONAL RESOURCES

- (20 min.) Have students do **Para hispanohablantes,** *TRB,* p. 206. ■
- (20 min.) Have students do **Para angloparlantes,** *TRB,* p. 206.

Practice Options/Homework Suggestions

- Internet (go.hrw.com, Keyword: WN3 PERSPECTIVAS-LEC)
- Have students do **Escritura creativa,** p. 397.
- Have students study **Vocabulario adicional, "El libro talonario,"** *TRB,* p. 302. ◆ ●
- *Advanced Placement Literature Preparation Book,* pp. 85–89 ▲

▲ = Advanced Learners ◆ = Slower Pace Learners ● = Special Learning Needs ■ = Heritage Speakers

Teacher's Name ______________________ Class ______________________ Date ______________

COLECCIÓN 6

Perspectivas humorísticas

DAY 11 50-MINUTE LESSON PLAN

STANDARDS FOR FOREIGN LANGUAGE LEARNING: DAY 11

Lectura

Communication 1.2: Students understand and interpret written and spoken language on a variety of topics.

Communication 1.3: Students present information, concepts, and ideas to an audience of listeners or readers on a variety of topics.

CORE INSTRUCTION

Warm-Up

- (5 min.) Have students study the expressions in **Así se dice,** p. 396.

Lectura

Teach

- (10 min.) Present **¿Te acuerdas?,** p. 396.
- (30 min.) In small groups, have students do **Cuaderno del escritor,** p. 397. See **Cuaderno del escritor,** *TRB,* p. 205.

Wrap-Up

- (5 min.) Have one student from each group give examples of their work from **Cuaderno del escritor.**

OPTIONAL RESOURCES

- (15 min.) Review the present perfect subjunctive with students, if necessary. See p. R58.

Practice Options/Homework Suggestions

- Internet (go.hrw.com, Keyword: WN3 PERSPECTIVAS-LEC)
- Have students do **Escritura informativa,** p. 397. See **Escritura informativa,** *TRB,* p. 206.

▲ = Advanced Learners ◆ = Slower Pace Learners ● = Special Learning Needs ■ = Heritage Speakers

Teacher's Name ______________________ Class ______________________ Date ____________

Perspectivas humorísticas

DAY 12 50-MINUTE LESSON PLAN

STANDARDS FOR FOREIGN LANGUAGE LEARNING: DAY 12

Vocabulario

Communication 1.1: Students engage in conversations, provide and obtain information, express feelings and emotions, and exchange opinions.

Communication 1.2: Students understand and interpret written and spoken language on a variety of topics.

Communities 5.1: Students use the language both within and beyond the school setting.

CORE INSTRUCTION

Warm-Up

- (5 min.) Have students review the main events of **"El libro talonario,"** pp. 391–395.

Vocabulario

Teach

- (15 min.) Have students do **Vocabulario en contexto,** Activity C, p. 406.
- (15 min.) Have students do Activity D, p. 406. You may wish to use Audio CD 2, Track 23. See script, *TRB,* p. 215.
- (10 min.) Have students do Activity E, p. 407.

Wrap-Up

- (5 min.) Have students review **Vocabulario esencial, "El libro talonario,"** p. 427.

OPTIONAL RESOURCES

- (10 min.) Have students do Additional Practice, *TRB,* p. 214. ◆ ●
- (30 min.) Have students do Group Work, *TRB,* p. 214.

Practice Options/Homework Suggestions

- *Cuaderno de práctica,* **Vocabulario adicional,** p. 152 ◆ ●
- Have students study for **Prueba de lectura.**

▲ = Advanced Learners ◆ = Slower Pace Learners ● = Special Learning Needs ■ = Heritage Speakers

Teacher's Name ______________________ Class ______________________ Date ____________

COLECCIÓN 6

Perspectivas humorísticas

DAY 13 50-MINUTE LESSON PLAN

STANDARDS FOR FOREIGN LANGUAGE LEARNING: DAY 13

Lectura

Communication 1.2: Students understand and interpret written and spoken language on a variety of topics.

Connections 3.1: Students reinforce and further their knowledge of other disciplines through the foreign language.

Connections 3.2: Students acquire information and recognize the distinctive viewpoints that are only available through the foreign language and its cultures.

CORE INSTRUCTION

Warm-Up

- (5 min.) Have students review **"El libro talonario,"** pp. 391–395.

Lectura

Assess

- (30 min.) Give **Prueba de lectura: "El libro talonario,"** *Assessment Program,* pp. 201–202.

Lectura

Teach

- (5 min.) Discuss the differences between a short story and a novel with students. Have students do Getting Started, *TRB,* p. 208.
- (5 min.) Discuss the various types of novels with students. See **Elementos de literatura,** p. 399.

Wrap-Up

- (5 min.) Have students give examples of each type of novel mentioned in **Elementos de literatura,** p. 399.

OPTIONAL RESOURCES

- (30 min.) Have students read **A leer por tu cuenta: "El soneto,"** and **Conoce al escritor,** pp. 402–403. See **"El soneto,"** *TRB,* pp. 211–212. ▲

Practice Options/Homework Suggestions

- Internet (go.hrw.com, Keyword: WN3 PERSPECTIVAS-LEC)
- Have students practice the Reading Strategy, *TRB,* p. 276. ◆ ●
- *Cuaderno de práctica,* pp. 115–117

▲ = Advanced Learners ◆ = Slower Pace Learners ● = Special Learning Needs ■ = Heritage Speakers

Teacher's Name ______________________ Class ______________________ Date ____________

Perspectivas humorísticas

DAY 14 50-MINUTE LESSON PLAN

STANDARDS FOR FOREIGN LANGUAGE LEARNING: DAY 14

Vocabulario

Communication 1.3: Students present information, concepts, and ideas to an audience of listeners or readers on a variety of topics.

Connections 3.1: Students reinforce and further their knowledge of other disciplines through the foreign language.

Comparisons 4.1: Students demonstrate understanding of the nature of language through comparisons of the language studied and their own.

Communities 5.1: Students use the language both within and beyond the school setting.

CORE INSTRUCTION

Warm-Up

- (5 min.) Have students review **Elementos de literatura: La novela,** p. 399.

Vocabulario

Teach

- (20 min.) Present **Mejora tu vocabulario: Las palabras cultas,** pp. 407–408. See **Mejora tu vocabulario: Las palabras cultas,** *TRB,* p. 215.
- (20 min.) Have students do Activity F, p. 408. See Activity F, *TRB,* p. 215.

Wrap-Up

- (5 min.) Have students do Critical Thinking, *TRB,* p. 215.

OPTIONAL RESOURCES

- (30 min.) Have students discuss Applying the Element, *TRB,* p. 208.
- (20 min.) Have students do Additional Practice, *TRB,* p. 209. ◆ ●

Practice Options/Homework Suggestions

- Have students do History Link, *TRB,* pp. 208–209.
- Have students do Community Link, *TRB,* p. 209.

Assessment Options

- *Assessment Program,* **Prueba de lectura: "El soneto,"** pp. 203–204
- *Assessment Program,* **Prueba de vocabulario adicional,** p. 275

▲ = Advanced Learners ◆ = Slower Pace Learners ● = Special Learning Needs ■ = Heritage Speakers

Teacher's Name ______________________ Class ____________________ Date __________

COLECCIÓN 6

Perspectivas humorísticas

DAY 15 50-MINUTE LESSON PLAN

STANDARDS FOR FOREIGN LANGUAGE LEARNING: DAY 15

Vocabulario

Communication 1.1: Students engage in conversations, provide and obtain information, express feelings and emotions, and exchange opinions.

Communication 1.2: Students understand and interpret written and spoken language on a variety of topics.

Communication 1.3: Students present information, concepts, and ideas to an audience of listeners or readers on a variety of topics.

Comparisons 4.1: Students demonstrate understanding of the nature of language through comparisons of the language studied and their own.

Communities 5.1: Students use the language both within and beyond the school setting.

CORE INSTRUCTION

Warm-Up

- (5 min.) Have students do Activity J, p. 409.

Vocabulario

Teach

- (20 min.) Have pairs of students do Activity G, pp. 408–409.
- (20 min.) Have students do Activity I, p. 409. Divide class into four groups and have each group do two words. See Activity I, *TRB,* p. 216.

Wrap-Up

- (5 min.) Have student volunteers from each group give roots and definitions for their words from Activity J.

OPTIONAL RESOURCES

- (30 min.) Have students do Activity H, p. 409. See Activity H, *TRB,* p. 215.
- (20 min.) Have students do **Para hispanohablantes,** *TRB,* p. 216. ■
- (10 min.) Have students do **Para angloparlantes,** *TRB,* p. 216.

Practice Options/Homework Suggestions

- Have students study **Vocabulario esencial, Mejora tu vocabulario,** p. 427. ◆ ●
- *Cuaderno de práctica,* p. 119
- Have students study for **Prueba de vocabulario.**
- *Advanced Placement Literature Preparation Book,* pp. 90–91 ▲

▲ = Advanced Learners ◆ = Slower Pace Learners ● = Special Learning Needs ■ = Heritage Speakers

Teacher's Name ______________________ Class ______________________ Date ______________

COLECCIÓN 6

Perspectivas humorísticas

DAY 16 50-MINUTE LESSON PLAN

STANDARDS FOR FOREIGN LANGUAGE LEARNING: DAY 16

Vocabulario/Gramática

Communication 1.1: Students engage in conversations, provide and obtain information, express feelings and emotions, and exchange opinions.

Communication 1.2: Students understand and interpret written and spoken language on a variety of topics.

Comparisons 4.1: Students demonstrate understanding of the nature of language through comparisons of the language studied and their own.

CORE INSTRUCTION

Warm-Up

- (5 min.) Have students review **Vocabulario,** pp. 405–409.

Vocabulario

Assess

- (20 min.) Give **Prueba de vocabulario,** *Assessment Program,* pp. 205–206.

Gramática

Teach

- (20 min.) Present **El aspecto progresivo,** pp. 413–414. See **El aspecto progresivo,** *TRB,* p. 218.

Wrap-Up

- (5 min.) Have students give examples of sentences using **ir, venir, andar, llevar,** and **pasar** with the progressive.

OPTIONAL RESOURCES

- (20 min.) Have students do **Ampliación, Hoja de práctica 6-A: Más sobre el progresivo,** *TRB,* p. 291. ◆ ●

Practice Options/Homework Suggestions

- *Cuaderno de práctica,* Activity 6, p. 123
- *Cuaderno de práctica,* **Ampliación, Hoja de práctica 6-A,** p. 153 ◆ ●

▲ = Advanced Learners ◆ = Slower Pace Learners ● = Special Learning Needs ■ = Heritage Speakers

Teacher's Name ______________________ Class ______________________ Date __________

COLECCIÓN 6

Perspectivas humorísticas

DAY 17 50-MINUTE LESSON PLAN

STANDARDS FOR FOREIGN LANGUAGE LEARNING: DAY 17

Gramática

Communication 1.2: Students understand and interpret written and spoken language on a variety of topics.

Communication 1.3: Students present information, concepts, and ideas to an audience of listeners or readers on a variety of topics.

Comparisons 4.1: Students demonstrate understanding of the nature of language through comparisons of the language studied and their own.

CORE INSTRUCTION

Warm-Up

- (5 min.) Have students review **El aspecto progresivo,** pp. 413–414.

Gramática

Teach

- (15 min.) Have students do Activity D, p. 414.
- (10 min.) Have students do Activity E, pp. 414–415.
- (15 min.) Have students do Activity F, p. 415.

Wrap-Up

- (5 min.) Have students read aloud examples from Activity H and I, p. 416.

OPTIONAL RESOURCES

- (30 min.) Have students do the second Pair Work, *TRB,* p. 218.
- (20 min.) Have students do Critical Thinking, *TRB,* p. 219. ▲
- (20 min.) Have students do Group Work, *TRB,* p. 219.

Practice Options/Homework Suggestions

- *Cuaderno de práctica,* Activities 7–11, pp. 123–125
- Have students do Activity G, H, or I, p. 416.
- Have students study for **Prueba de gramática**.

Assessment Options

- *Assessment Program,* **Prueba de gramática 6-A: Más sobre el progresivo,** p. 276

▲ = Advanced Learners ◆ = Slower Pace Learners ● = Special Learning Needs ■ = Heritage Speakers

Perspectivas humorísticas

DAY 18 50-MINUTE LESSON PLAN

STANDARDS FOR FOREIGN LANGUAGE LEARNING: DAY 18

Gramática

Communication 1.2: Students understand and interpret written and spoken language on a variety of topics.

Comparisons 4.1: Students demonstrate understanding of the nature of language through comparisons of the language studied and their own.

CORE INSTRUCTION

Warm-Up

- (5 min.) Have students review **El aspecto progresivo,** pp. 413–416.

Gramática

Assess

- (40 min.) Give **Prueba de gramática: El aspecto progresivo,** *Assessment Program,* pp. 208–209.

Wrap-Up

- (5 min.) Go over items on the quiz with students.

OPTIONAL RESOURCES

- (30 min.) Have students look up the same news story in English and Spanish on the Internet. Review translation of information.

Practice Options/Homework Suggestions

- Have students read **Comparación y contraste,** p. 417.
- *Cuaderno de práctica,* Activities 13–15, p. 127

▲ = Advanced Learners ◆ = Slower Pace Learners ● = Special Learning Needs ■ = Heritage Speakers

Teacher's Name ______________________ Class ______________________ Date __________

COLECCIÓN 6

Perspectivas humorísticas

DAY 19 50-MINUTE LESSON PLAN

STANDARDS FOR FOREIGN LANGUAGE LEARNING: DAY 19

Gramática

Communication 1.2: Students understand and interpret written and spoken language on a variety of topics.

Communication 1.3: Students present information, concepts, and ideas to an audience of listeners or readers on a variety of topics.

Comparisons 4.1: Students demonstrate understanding of the nature of language through comparisons of the language studied and their own.

CORE INSTRUCTION

Warm-Up

- (5 min.) Have students read **Comparación y contraste,** p. 417.

Gramática

Teach

- (15 min.) Present **Comparación y contraste,** p. 417. See **Comparación y contraste,** *TRB,* p. 219.
- (10 min.) Have students do Activity A, p. 417. See Activity A, *TRB,* pp. 219–220.
- (10 min.) Have students do Activity B, p. 417. See Activity B, *TRB,* p. 220.

Wrap-Up

- (10 min.) Have students do **Para angloparlantes,** *TRB,* p. 220.

OPTIONAL RESOURCES

- (30 min.) Have students do **Para hispanohablantes,** *TRB,* p. 220. ■

Practice Options/Homework Suggestions

- *Cuaderno de práctica,* Activity 12, p. 126
- Have students read **Panorama cultural,** pp. 400–401.
- Have students study for **Prueba de comparación y contraste.**

▲ = Advanced Learners ◆ = Slower Pace Learners ● = Special Learning Needs ■ = Heritage Speakers

Teacher's Name ______________________ Class ______________________ Date ____________

Perspectivas humorísticas

DAY 20 50-MINUTE LESSON PLAN

STANDARDS FOR FOREIGN LANGUAGE LEARNING: DAY 20

Gramática/Panorama cultural

Communication 1.1: Students engage in conversations, provide and obtain information, express feelings and emotions, and exchange opinions.

Communication 1.2: Students understand and interpret written and spoken language on a variety of topics.

Cultures 2.1: Students demonstrate an understanding of the relationship between the practices and perspectives of the culture studied.

Connections 3.2: Students acquire information and recognize the distinctive viewpoints that are only available through the foreign language and its cultures.

CORE INSTRUCTION

Warm-Up

- (5 min.) Have students review **Comparación y contraste,** p. 417.

Gramática

Assess

- (25 min.) Give **Prueba de comparación y contraste,** *Assessment Program,* p. 210.

Panorama cultural

Teach

- (5 min.) Have students discuss the introduction of **Panorama cultural,** p. 400. See Summary, **Panorama cultural,** *TRB,* p. 210.
- (10 min.) Play Audio CD 2, Tracks 20–21, and have students listen to the interviews.

Wrap-Up

- (5 min.) Have students do Presentation, *TRB,* p. 210.

OPTIONAL RESOURCES

- (20 min.) Have students do **Panorama cultural,** Teaching Suggestions, **Antes de ver,** *Video Guide,* p. 49.
- (15 min.) Show **Panorama cultural,** *Video Program* (Videocassette 2). See **Panorama cultural,** Teaching Suggestions, **Mientras lo ves,** *Video Guide,* p. 49.
- (20 min.) Have students do **Panorama cultural,** Teaching Suggestions, **Después de ver,** *Video Guide,* p. 49.
- (15 min.) Have students do **Hoja de actividades 3,** *Video Guide,* p. 53.

Practice Options/Homework Suggestions

- Have students do Activities A and B, p. 401.
- *Advanced Placement Literature Preparation Book,* pp. 132–134 ▲

▲ = Advanced Learners ◆ = Slower Pace Learners ● = Special Learning Needs ■ = Heritage Speakers

Teacher's Name ______________________ Class ______________________ Date ____________

COLECCIÓN **6**

Perspectivas humorísticas

DAY 21 50-MINUTE LESSON PLAN

STANDARDS FOR FOREIGN LANGUAGE LEARNING: DAY 21

Panorama cultural/Comunidad y oficio

Communication 1.1: Students engage in conversations, provide and obtain information, express feelings and emotions, and exchange opinions.

Communication 1.2: Students understand and interpret written and spoken language on a variety of topics.

Connections 3.1: Students reinforce and further their knowledge of other disciplines through the foreign language.

Connections 3.2: Students acquire information and recognize the distinctive viewpoints that are only available through the foreign language and its cultures.

Comparisons 4.1: Students demonstrate understanding of the nature of language through comparisons of the language studied and their own.

Communities 5.1: Students use the language both within and beyond the school setting.

Communities 5.2: Students show evidence of becoming lifelong learners by using the language for personal enjoyment and enrichment.

CORE INSTRUCTION

Warm-Up

- (5 min.) Have students review the interviews on pp. 400–401.

Panorama cultural

Teach

- (15 min.) In pairs, have students do Activity C, p. 401.

Comunidad y oficio

Teach

- (5 min.) Present **Comunidad y oficio,** p. 404. See Getting Started, *TRB,* p. 213.
- (10 min.) Have students read aloud **Comunidad y oficio,** p. 404.
- (10 min.) Show **Comunidad y oficio,** *Video Program* (Videocassette 2). See **Comunidad y oficio,** Teaching Suggestions, **Mientras lo ves,** *Video Guide,* p. 50.

Wrap-Up

- (5 min.) Have students do **Hoja de actividades 4, Mientras lo ves,** *Video Guide,* p. 54.

OPTIONAL RESOURCES

- (10 min.) Have students do Activity D, p. 401. You may use Audio CD 2, Track 22. See script, *TRB,* p. 210.
- (30 min.) Have students do Pair Work, *TRB,* p. 213.
- (15 min.) Have students do the first Thinking Critically activity, *TRB,* p. 213. ▲
- (15 min.) Have students do Language Note, *TRB,* p. 213.
- (10 min.) Have students do **Comunidad y oficio,** Teaching Suggestions, **Antes de ver,** *Video Guide,* p. 50.
- Have students do the second Thinking Critically activity, *TRB,* p. 213. ▲
- (20 min.) Have students do **Investigaciones,** Activity B, p. 404. See the third Thinking Critically activity, *TRB,* p. 213. ▲

Practice Options/Homework Suggestions

- Internet (go.hrw.com, Keyword: WN3 PERSPECTIVAS-CYO)
- Have students do **Investigaciones,** Activity A, p. 404.

▲ = Advanced Learners ◆ = Slower Pace Learners ● = Special Learning Needs ■ = Heritage Speakers

Teacher's Name ______________ Class ______________ Date ______________

Perspectivas humorísticas

DAY 22 50-MINUTE LESSON PLAN

STANDARDS FOR FOREIGN LANGUAGE LEARNING: DAY 22

Ortografía

Communication 1.1: Students engage in conversations, provide and obtain information, express feelings and emotions, and exchange opinions.

Connections 3.1: Students reinforce and further their knowledge of other disciplines through the foreign language.

Comparisons 4.1: Students demonstrate understanding of the nature of language through comparisons of the language studied and their own.

CORE INSTRUCTION

Warm-Up

- (5 min.) Have students read the introduction of **Los verbos que terminan en *-ear*,** p. 418.

Ortografía

Teach

- (10 min.) Present **Los verbos que terminan en *-ear*,** p. 418. See **Los verbos que terminan en *-ear*,** *TRB,* p. 221.
- (5 min.) Have students do Activity A, p. 418.
- (10 min.) Present **La acentuación,** p. 419. See **La acentuación,** *TRB,* pp. 221–222.
- (15 min.) Have students do Activity C, p. 419.

Wrap-Up

- (5 min.) Have students review **¡Ojo!,** p. 418. See **¡Ojo!,** *TRB,* p. 221.

OPTIONAL RESOURCES

- (10 min.) Have students do Cooperative Learning, *TRB,* p. 221. ▲
- (10 min.) Have students do Pair Work, *TRB,* p. 221.
- (10 min.) Have students do Pair Work, *TRB,* p. 222.
- (10 min.) Have students do Additional Practice, *TRB,* p. 222. ◆ ●
- (20 min.) Have students do Group Work, *TRB,* p. 222. ◆ ●
- (30 min.) Have students do the second Additional Practice, *TRB,* p. 222. ◆ ●
- (30 min.) Have students do Challenge, *TRB,* p. 222. ▲

Practice Options/Homework Suggestions

- Have students do Activity B, p. 418.
- Have students do Activity D, p. 419.
- *Cuaderno de práctica,* Activities 1–6, pp. 128–129
- Have students study for **Prueba de ortografía.**

▲ = Advanced Learners ◆ = Slower Pace Learners ● = Special Learning Needs ■ = Heritage Speakers

Teacher's Name ______________________ Class ______________________ Date ______________

Perspectivas humorísticas

DAY 23 50-MINUTE LESSON PLAN

STANDARDS FOR FOREIGN LANGUAGE LEARNING: DAY 23

Ortografía

Communication 1.2: Students understand and interpret written and spoken language on a variety of topics.

Communication 1.3: Students present information, concepts, and ideas to an audience of listeners or readers on a variety of topics.

Comparisons 4.1: Students demonstrate understanding of the nature of language through comparisons of the language studied and their own.

Communities 5.1: Students use the language both within and beyond the school setting.

CORE INSTRUCTION

Warm-Up

- (5 min.) Have students review **Ortografía,** pp. 418–419.

Ortografía
Teach

- (20 min.) Give **Dictado,** Activities A and B, p. 419. You may wish to use Audio CD 2, Tracks 24–25. See scripts, *TRB,* pp. 222–223.

Ortografía
Assess

- (20 min.) Give **Prueba de ortografía,** *Assessment Program,* p. 212.

Wrap-Up

- (5 min.) Review items on quiz with students.

OPTIONAL RESOURCES

- (15 min.) Have students do the second Group Work, *TRB,* p. 222.
- (20 min.) Have students do **Para hispanohablantes,** *TRB,* p. 223. ■
- (20 min.) Have students do **Para angloparlantes,** *TRB,* p. 223.

Practice Options/Homework Suggestions

- Have students read **Taller del escritor,** p. 420.

▲ = Advanced Learners ◆ = Slower Pace Learners ● = Special Learning Needs ■ = Heritage Speakers

Perspectivas humorísticas

DAY 24 50-MINUTE LESSON PLAN

STANDARDS FOR FOREIGN LANGUAGE LEARNING: DAY 24

Taller del escritor

Communication 1.1: Students engage in conversations, provide and obtain information, express feelings and emotions, and exchange opinions.

Communication 1.2: Students understand and interpret written and spoken language on a variety of topics.

Communication 1.3: Students present information, concepts, and ideas to an audience of listeners or readers on a variety of topics.

Connections 3.1: Students reinforce and further their knowledge of other disciplines through the foreign language.

Communities 5.1: Students use the language both within and beyond the school setting.

CORE INSTRUCTION

Warm-Up

- (5 min.) Have students review their work on the portfolio suggestions after each reading selection. See **Antes de escribir: Cuaderno del escritor,** p. 420.

Taller del escritor

Teach

- (5 min.) Introduce **Taller del escritor,** p. 420. See Presenting the Workshop, *TRB,* p. 224.
- (10 min.) Have students read **Especulación sobre causas y efectos** and **Preguntas y escritura libre,** p. 420.
- (15 min.) Have students begin freewriting. They may use the topic and questions in **Escritura libre,** p. 420.
- (10 min.) Present **Explora causas y efectos,** p. 421. Go over **Diagrama de efectos positivos y negativos** with students, p. 421.

Wrap-Up

- (5 min.) Discuss **Recopila pruebas,** p. 421, with students and **Antes de escribir,** *TRB,* p. 224.

OPTIONAL RESOURCES

- (20 min.) Have students read and discuss **Diagrama de efectos positivos y negativos,** p. 421.
- (20 min.) Have students do **Investiga los medios de comunicación,** p. 421.

Practice Options/Homework Suggestions

- Have students choose the theme of their essay.

▲ = Advanced Learners ◆ = Slower Pace Learners ● = Special Learning Needs ■ = Heritage Speakers

Teacher's Name ______________ Class ______________ Date ______________

COLECCIÓN 6

Perspectivas humorísticas

DAY 25 50-MINUTE LESSON PLAN

STANDARDS FOR FOREIGN LANGUAGE LEARNING: DAY 25

Taller del escritor

Communication 1.2: Students understand and interpret written and spoken language on a variety of topics.

Connections 3.1: Students reinforce and further their knowledge of other disciplines through the foreign language.

Communities 5.1: Students use the language both within and beyond the school setting.

CORE INSTRUCTION

Warm-Up

- (5 min.) Have students read **El borrador: Organización,** p. 422.

Taller del escritor

Teach

- (10 min.) Review the different parts of the essay discussed in **Organización** with students. See **El borrador,** *TRB,* p. 224.
- (10 min.) Present **Desarrolla tu estilo: Tipos de oraciones** and **Relaciona ideas,** p. 422.
- (20 min.) Have students work on their essays.

Wrap-Up

- (5 min.) Have students review **Esquema para el ensayo de especulación sobre causas o efectos,** p. 422.

OPTIONAL RESOURCES

- (15 min.) Have students find examples of cause-and-effect writing in newspapers.
- (20 min.) Have students compare the authors in this collection and their works, pp. 383, 395, and 403.

Practice Options/Homework Suggestions

- Have students work on their drafts.

▲ = Advanced Learners ◆ = Slower Pace Learners ● = Special Learning Needs ■ = Heritage Speakers

Teacher's Name ______________________ Class ______________________ Date ____________

COLECCIÓN 6

Perspectivas humorísticas

DAY 26 50-MINUTE LESSON PLAN

STANDARDS FOR FOREIGN LANGUAGE LEARNING: DAY 26

Taller del escritor

Communication 1.3: Students present information, concepts, and ideas to an audience of listeners or readers on a variety of topics.

Connections 3.1: Students reinforce and further their knowledge of other disciplines through the foreign language.

CORE INSTRUCTION

Warm-Up

- (5 min.) Have students read **Así se dice,** p. 422, and **Pautas para redactar,** p. 423.

Taller del escritor

Teach

- (30 min.) In small groups, have students do **Evaluación y revisión: Intercambio entre compañeros,** p. 422. Have them use expressions from **Así se dice,** p. 422.
- (10 min.) Have students evaluate their own work using the **Pautas de evaluación** and **Técnicas de revisión,** p. 423.

Wrap-Up

- (5 min.) Have student volunteers give examples of the feedback they gave their partner or that they themselves received.

OPTIONAL RESOURCES

- (20 min.) Have students do Reteaching, *TRB,* p. 225. ◆ ●

Practice Options/Homework Suggestion

- Have students revise their drafts.

▲ = Advanced Learners ◆ = Slower Pace Learners ● = Special Learning Needs ■ = Heritage Speakers

Teacher's Name ______________ Class ______________ Date ______________

COLECCIÓN 6

Perspectivas humorísticas

DAY 27 50-MINUTE LESSON PLAN

STANDARDS FOR FOREIGN LANGUAGE LEARNING: DAY 27

Taller del escritor

Communication 1.1: Students engage in conversations, provide and obtain information, express feelings and emotions, and exchange opinions.

Communication 1.3: Students present information, concepts, and ideas to an audience of listeners or readers on a variety of topics.

Connections 3.1: Students reinforce and further their knowledge of other disciplines through the foreign language.

CORE INSTRUCTION

Warm-Up

- (5 min.) Have students exchange their drafts and proofread for spelling.

Taller del escritor

Teach

- (20 min.) Have students do **Corrección de pruebas,** p. 424. See suggestions in **Corrección de pruebas,** *TRB,* p. 225.
- (15 min.) Have students do one of the activities in **Publicación,** p. 424.
- (5 min.) Have students do **Reflexión,** p. 424. See **Reflexión,** *TRB,* p. 225. Have them use the expressions in **Así se dice,** p. 424.

Wrap-Up

- (5 min.) Have students do Closure, *TRB,* p. 225.

OPTIONAL RESOURCES

- (20 min.) Have students do **Publicación,** *TRB,* p. 225. ▲
- (20 min.) Have students follow the Assessment criteria, *TRB,* p. 225.

Practice Options/Homework Suggestions

- Have students do **A ver si puedo…,** pp. 425–426.

▲ = Advanced Learners ◆ = Slower Pace Learners ● = Special Learning Needs ■ = Heritage Speakers

Teacher's Name ______________________ Class ______________________ Date ______________

COLECCIÓN 6

Perspectivas humorísticas

DAY 28 50-MINUTE LESSON PLAN

STANDARDS FOR FOREIGN LANGUAGE LEARNING: DAY 28

A ver si puedo…

Communication 1.1: Students engage in conversations, provide and obtain information, express feelings and emotions, and exchange opinions.

Communication 1.2: Students understand and interpret written and spoken language on a variety of topics.

Communication 1.3: Students present information, concepts, and ideas to an audience of listeners or readers on a variety of topics.

Connections 3.1: Students reinforce and further their knowledge of other disciplines through the foreign language.

Connections 3.2: Students acquire information and recognize the distinctive viewpoints that are only available through the foreign language and its cultures.

Communities 5.1: Students use the language both within and beyond the school setting.

CORE INSTRUCTION

Warm-Up

- (5 min.) Have students review the objectives listed on the Collection Opener, p. 376.

A ver si puedo…

Review

- (10 min.) Have students do **Lectura,** Activities A and B, p. 425.
- (10 min.) Have students do **Cultura,** Activity C, p. 425.
- (10 min.) Have students do **Comunicación,** Activities D, E, F, and G, pp. 425–426.
- (10 min.) Have students do **Escritura,** Activities H, I, and J, p. 426.

Wrap-Up

- (5 min.) Answer any questions about either of the two chapter exams.

OPTIONAL RESOURCES

- (35 min.) Have students read **Enlaces literarios: El teatro latinoamericano del siglo XX,** pp. 428–438. See *TRB,* pp. 228–231. ▲
- (20 min.) Have students do **Comprensión del texto,** p. 439. See **Comprensión del texto,** *TRB,* pp. 229–230. ▲
- (25 min.) Have students do **Análisis del texto,**p. 439. See **Análisis del texto,** *TRB,* p. 230–231. ▲
- (25 min.) Have students do **Más allá del texto,** p. 439. See **Más allá del texto,** *TRB,* p. 231. ▲

Practice Options/Homework Suggestions

- Have students study for the **Examen de lengua.**

▲ = Advanced Learners ◆ = Slower Pace Learners ● = Special Learning Needs ■ = Heritage Speakers

Teacher's Name ______________________ Class ______________________ Date ____________

COLECCIÓN 6

Perspectivas humorísticas

DAY 29 50-MINUTE LESSON PLAN

CORE INSTRUCTION

Assess

- (50 min.) Give **Colección 6 Examen de lengua,** *Assessment Program,* pp. 219–225.

OPTIONAL RESOURCES

- (50 min.) Give **Examen de lectura: de *Don Quijote de la Mancha* y "El libro talonario,"** *Assessment Program,* pp. 213–218. To allow students more time to take the exam, either **Examen** may be given over two class periods.

Practice Options/Homework Suggestions

- Have students study for the **Examen de lectura: de *Don Quijote de la Mancha* y "El libro talonario."**

Assessment Options

- *Assessment Program,* Performance Assessment, p. 300
- *Assessment Program,* Final Exam, **Examen Final: Lectura,** pp. 233–235
- *Assessment Program,* Final Exam, **Examen Final: Lengua,** pp. 236–240

▲ = Advanced Learners ◆ = Slower Pace Learners ● = Special Learning Needs ■ = Heritage Speakers

Perspectivas humorísticas

DAY 30 50-MINUTE LESSON PLAN

CORE INSTRUCTION

Assess

- (50 min.) Give **Examen de lectura: de *Don Quijote de la Mancha* y "El libro talonario,"** *Assessment Program,* pp. 213–218.

OPTIONAL RESOURCES

- (50 min.) Give **Colección 6 Examen de lengua,** *Assessment Program,* pp. 219–225. To allow students more time to take the exam, either **Examen** may be given over two class periods.

Practice Options/Homework Suggestions

- Internet (go.hrw.com, Keyword: WN3 PERSPECTIVAS)

Assessment Options

- *Assessment Program,* Performance Assessment, p. 300
- *Assessment Program,* Final **Exam, Examen Final: Lectura,** pp. 233–235
- *Assessment Program,* Final **Exam, Examen Final: Lengua,** pp. 236–240

▲ = Advanced Learners ◆ = Slower Pace Learners ● = Special Learning Needs ■ = Heritage Speakers